The Pelton Water Wheel System of Power

by Pelton Water Wheel Company

with an introduction by Roger Chambers

Self Reliance Books

Get more historic titles on animal and stock breeding, gardening and old fashioned skills by visiting us at:

http://selfreliancebooks.blogspot.com/

Introduction

I am pleased to present yet another title on Homesteading and Farm Life.

This volume is entitled "The Pelton System of Power" and was published in 1909.

The work is in the Public Domain and is re-printed here in accordance with Federal Laws.

As with all reprinted books of this age that are intended to perfectly reproduce the original edition, considerable pains and effort had to be undertaken to correct fading and sometimes outright damage to existing proofs of this title. At times, this task is quite monumental, requiring an almost total "rebuilding" of some pages from digital proofs of multiple copies. Despite this, imperfections still sometimes exist in the final proof and may detract from the visual appearance of the text.

I hope you enjoy reading this book as much as I enjoyed making it available to readers again.

Roger Chambers

Fig. 11—C. E. COLBURN'S FARM AND STOCK BARN

Fig. 19—MR. LAWSON VALENTINE'S BARN, "HOUGHTON FARM," MOUNTAINVILLE, N. Y.

2

Foreword

HYDRAULICS, in its application to water wheels, is now considered as an exact science.

The impulse wheel, of which the PELTON is regarded as the *type*, was primarily the result of the accidental observance of water action on moving vanes. Since then its development and improvement have been based on scientific and mathematical deductions—the result being the PELTON System of Power, with which this catalog deals.

An endeavor is herein made to explain the primary principles and functions of the PELTON WHEEL, as well as to show the facility with which it may be adapted to a great variety of conditions. To this end illustrations have been freely employed, all of which are from photographs of machinery actually constructed.

No claims are made that cannot be fully substantiated in actual practise. An experience of twenty years in the design and construction of water wheel apparatus, together with the actual manufacture of *twelve thousand six hundred* wheels during that period, is the assurance offered by this Company of its ability to successfully cope with any proposition within its sphere.

General Information

The PELTON WATER WHEEL is the pioneer of what is known as the "impulse and reaction" type, as distinct from the turbine wheel. It is essentially a "high head" wheel—that is to say, is better adapted for operating under high pressure (twenty-five feet and over), and utilizing a small quantity of water—relying on the impulsive force of the water on moving vanes, rather than the constant and uniform pressure of a large, slow-moving mass of water.

In simplest form it consists of a cast-iron or steel center, to the periphery of which are attached cups, or "buckets," as they are technically called. The high efficiency of the PELTON WHEEL is largely due to the design and construction of the buckets, which are fully covered by patents. These receive the stream at a tangent to the periphery of the wheel, absorb the force or power, and discharge the water at practically no velocity, and at a slight angle, so as to avoid interference with the succeeding bucket.

The wheel is carried usually on a horizontal shaft, supported by journal boxes, but in special cases the shaft may be vertical with step-and-thrust bearing. The water, being led to the wheel by means of a pipe, impinges on the buckets through a nozzle, the end of which is fitted with a cylindrical tip of diameter proportioned to the head of water and amount of power to be developed. Different diameters of tips can be screwed into the nozzle, thereby varying the power of the wheel from the maximum, which is limited by the size of the bucket, down to a small proportion of the rated capacity; in this way no more water is used than is actually required for the power, and a uniformly high efficiency is maintained at all stages of load.

The power is transmitted from the wheel shaft, either by pulley and belt, or by direct-connecting the wheel shaft and that of the machinery to be driven by means of a coupling.

Power and Adaptation

The power of the PELTON WHEEL does not depend on its diameter, but upon the head and amount of water applied to it — in other words, the size of nozzle used — which establishes the size of bucket on the wheel. The diameter of the wheel determines its speed, and, under a given head, the number of revolutions at which a certain size of wheel runs should be constant, irrespective of the power developed.

Where a very considerable amount of power is required under a comparatively low head, a wheel of large diameter is sometimes necessary, to admit of buckets of the proper size, as also the application of two or more nozzles for the purpose of multiplying the power.

Where wheels of standard sizes do not meet the requirements of any particular case, special wheels are often made, of such diameter as will give the proper speed, with buckets to suit the power conditions. In some instances they are required to be as large as 20 to 35 feet in diameter, to conform to the machinery to be operated — admitting in this way of direct-connection to crank-shafts of pumps, compressors or other slow-speed machinery. The facility with which such adaptation can be made to all varying conditions, is one of the marked and distinguishing features of the PELTON WHEEL, and admits of application to every possible service in the most economic and efficient way.

It will thus be seen that the PELTON WHEEL is a very simple piece of apparatus, involving the use of but little mechanical construction, and whether the power be one horse-power or ten thousand horse-power, the same elements are involved. As a consequence, the PELTON System of Power possesses primarily the great advantages of simplicity of construction and absence of wearing parts, insuring a prime mover of absolute reliability.

Use of Double Nozzles

Where the speed requirements call for a wheel of small diameter, on which it is not possible to obtain sufficient power from one nozzle, a double nozzle is used, which allows two streams of water to impinge on the same wheel; the power of the wheel is thus doubled, necessarily using water in like proportion.

This arrangement is particularly adapted for running electric machinery by means of direct-connection of water wheel and generator shafts, and is applicable to any high-speed machinery. The advantage gained by the absence of all belting or gearing, with its attendant loss of power and cost of maintenance, will be readily appreciated. Numerous illustrations of these applications are given on succeeding pages.

Conditions as to Head

Experiments have shown that the PELTON WHEEL will give as high an efficiency as any form of turbine under heads as low as from 10 to 12 feet, but its construction does not admit of handling sufficient water to develop any considerable amount of power under so low a head within a reasonable limit of cost. It is not, therefore, recommended for heads of less than about 20 feet, except where but a comparatively small amount of power is required.

The PELTON-FRANCIS turbine, described on page 102, is designed to meet such low head conditions as do not favor the PELTON type of wheel.

Under low or medium heads, where the capacity of a wheel is not sufficient even with a double nozzle, two or more wheels may be mounted on the same shaft, their nozzles being connected to the main pipe by means of a "Y" or receiver—the power obtained in the unit thus being in direct proportion to the number of wheels used.

As regards extreme head, there is practically no limit under which the PELTON WHEEL may be safely and efficiently operated. Heads of from 1200 to 1600 feet are of common occurrence, and PELTON WHEELS are now running under heads as high as 2150 feet.

Conservation of Head

As water, in most cases where available for power, has a positive commercial value, the most advantageous and profitable use of it should be considered. In this relation not only the wheel, but the pipe which supplies the motive power, has an important bearing upon the result.   The loss of head by the friction of water in its passage through a line of pipe is more serious than is generally supposed, and must be carefully estimated in all cases where water is conveyed in this way any considerable distance.

Head, in this connection, is simply a convertible term for power, and has a direct relation to value.   It should, therefore, be conserved to as great an extent as the circumstances of the case will justify.   Where both the water supply and head are limited, the pipe should be of sufficient capacity to avoid loss of head as far as possible. Where water is abundant and a very considerable head can be obtained, a greater loss in pipe friction may be justified to save cost in pipe.

In order to obtain the most economic results, each individual case should be considered in view of the existing conditions, having particular reference to the advantage of additional power, cost of pipe, and the fact that provision in pipe line for an ultimate power beyond present needs is often advisable.

Advantages of a High Head

The principle upon which the PELTON WHEEL operates being that of direct pressure, makes a high head desirable when it can be obtained at a reasonable cost. The amount of water required to develop a given power decreases in direct ratio as the head or pressure increases—consequently the additional length of pipe necessary in such case is often compensated for by its reduced size and thickness; also the same result may be obtained with less water and a smaller wheel.   The lessened expense in this way frequently justifies a largely increased expenditure in pipe line for the purpose of securing a higher head—especially as a small amount of water can, by such means, be made serviceable for many purposes.   The conditions presented in each particular case must, however, determine in regard to this.

Durability and Reliability

These features, next to efficiency, are conceded to be of the first importance in machinery of this character when used for any purpose, and more especially when operating electric plants in which water power is now so important a factor.   It should be noticed that in the PELTON WHEEL the wearing parts are reduced to a minimum, and that, in comparison with other forms of wheels, the mechanism is of the simplest kind—thus insuring freedom from accident and absolute reliability of service.   These points will especially appeal to those running power plants, in which continuous service is an absolute essential, and also to those who are located in foreign countries, to whom, in most cases, a breakdown would involve serious delay and loss.

Many wheels can be referred to that have been running continuously for about twenty years without any appreciable expense and without impairment of efficiency. Water carrying sand and grit, so destructive to other forms of wheels, has no effect upon the PELTON WHEEL except under extreme conditions, where a set of buckets may occasionally be required, involving but a small expense.

For Mill and Mine Work

The modern and most approved method of operating machinery of this character is that of having separate wheels for the various departments of work, such as batteries, crushers, concentrators, hoists, etc. By this means the different classes of machinery are under separate control, rendering the use of a governor unnecessary, and eliminating in large part all intermediate gearing.

The cost of additional wheels for such a distribution of power is generally more than compensated for by the saving in governor and connections referred to, as also by the convenience and flexibility of the plant when so arranged.

Where the water supply is limited, the wheel running the crusher may be set high enough to use the discharge for the batteries.

It often happens that although there is no water supply available for general power purposes, the drain water from the tunnel, if it is so situated as to obtain a fall, may be utilized on a small Pelton Wheel for operating an electric generator for lighting the mine and driving blowers.

For Electric Generators

The Pelton Wheel is most readily adapted for driving electric generators—its diameter being varied to suit the speed of the electric machinery, and the size and number of streams arranged for the output of power required. In all ordinary cases direct connection of wheel and generator shafts can be accomplished, thus eliminating the losses occasioned by belting and gearing, so frequently found in steam and turbine installations. The expense of floor and power house space required for these methods of transmission is also saved, thus effecting a double economy.

In choosing apparatus for electric transmission it is usually advisable to determine, within certain limits, the speeds and capacities to which the water wheels will be adapted without excessive expense, at the same time obtaining the highest efficiency. The generators may then be chosen between these limits, and the wheels closely designed to adapt them to the generators. This Company is always glad to co-operate with the electrical manufacturers with a view to obtaining the best combination of water wheels and generators for any proposed installation.

There are various ways in which water wheel units may be constructed for electric plants—the principal ones being as follows:

Iron-mounted Type: In this case the wheel is completely enclosed in sheet steel or cast-iron upper and lower housings, supported on a suitable cast-iron bed-plate, which also supports the pedestal bearings, nozzle and gate valve. All the parts are assembled in one complete construction and erected in the shops before shipment. The design is made to harmonize with that of the generator—maintaining the same proportion of bed-plate, pedestal bearings, and general appearance throughout. The use of this type of construction greatly simplifies the foundations and work of erecting, it merely being necessary to set the water wheel on the masonry floor and anchor it securely.

SEMI-MASONRY MOUNTED: In this style of construction the lower wheel housing and bed-plate are eliminated to a large extent, thus greatly reducing the weight; the bed-plate is replaced by a cast-iron frame securely anchored in the masonry foundation, which is carried up, sometimes above the power house floor level, to support it. The bed frame is fitted with the housing and pedestal bearings, also with the nozzle, gate valve, etc. The nozzle and all water-pressure connections come within the nozzle pit, which is an extension of the tail-race. A cast-iron grilled floor plate set in a cast-iron frame permits access to the nozzle pit—thus bringing the nozzle and connections all within reach of the crane.

Still a third method might be mentioned, which is only applicable in certain instances : this is to mount the water wheel on the shaft and in the pulley compartment of a dynamo arranged for belt driving. In this case the dynamo bed-plate must be provided with the necessary connecting flanges and be of suitable dimensions for receiving the water wheel parts.

Means of Utilizing Water Power

The idea of a water power is generally associated with a river or large stream, an expensive dam, huge flume, heavy grading and stone work, massive turbine wheels, pits, curbing and penstock—all involving so large an outlay as to make a power so produced of doubtful advantage, especially when so much is sacrificed in location of works as is generally necessary in such cases.

By means of the PELTON system only a small diverting dam is required, then a pipe running along the surface of the ground to the power station, which may be located at any convenient point and high enough to be out of the reach of floods. It is no exaggeration to say that, in this way, a small trout brook with a high head will often furnish as much power as a large stream under a low head, in a much more convertible form and at probably not more than one-fourth the outlay.

Notes in Regard to Wheel Mounting

For mining purposes, wheels are generally mounted on timber frame-work, as illustrated on page 24. In most cases it is advisable to build this on the ground, rather than pay freight for a considerable distance. Any millwright, or one reasonably familiar with such work, however, will have no difficulty in constructing the frame from such detailed drawings as are furnished.

If suitable timber is not available at the site, the frame is furnished when desired — in which case the wheel is mounted on frame, all pieces marked, then taken down and packed for shipment; in this way the parts can be readily assembled on the ground. In special cases, where large amounts of power are involved, it is often desirable to support the journal boxes by concrete piers and enclose the upper part of the wheel in a sheet iron housing.

On page 2 is shown the standard type of the iron-mounted PELTON WHEEL, which is more often used in connection with electrical apparatus, or such other machinery as requires a compact and self-contained driving power. The illustration shown represents a wheel with extended shaft and outboard bearing for driving by means of belt connection, but the same general mounting obtains in the event of its being intended to direct-connect water wheel and generator shafts. The iron-mounted type usually embraces automatic ring-oiling journal stands of design similar to those on generator, so as to present a harmonious appearance.

A special feature embraced in PELTON design is the CENTRIFUGAL DISC, which prevents leakage of water along the shaft and into the journals. This device, covered by PELTON patents, consists of a cast-iron disc attached to the wheel shaft and rotating within wall chambers, secured to the inside of the wheel housing. Any water collecting on the shaft is caught by the disc and thrown by centrifugal force into the chamber, whence it is drained through a tube into the tail-race. This frictionless device supersedes the old form of stuffing gland which required constant packing, and which, on account of grit, was often the cause of cutting or wearing the shaft.

It is impossible to designate all of the various mountings to which the PELTON WHEEL is adapted, as each wheel is designed to meet a particular case to the best advantage, but the illustrations on succeeding pages will indicate the possibilities in this regard.

Construction plans are furnished with all wheels, so that any competent mechanic can install them without difficulty.

Wide Range of Variation

One of the most notable advantages of the PELTON system is its adaptability to widely different conditions of water supply and power. By changing the nozzle-tip,  thereby varying the size of the stream thrown on the wheel, the power produced may be reduced from the maximum down to about twenty-five per cent of same, using water practically in direct proportion to the amount of power that is being developed, thus operating the wheel up to its full capacity with an ample water supply, or to the same relative advantage with a reduced quantity, when, for any reason, the supply fails in part.

Also, with a double nozzle wheel, one nozzle may be closed entirely and a small tip used on the other, admitting of a still greater reduction in capacity, while maintaining the same high efficiency.

The advantage of such means of adjustment to varying conditions will be apparent, as there are few cases where the water supply is not reduced at certain seasons of the year, making its economical and efficient use at such times most desirable. This arrangement also admits of using, without disadvantage, a wheel of larger capacity than present requirements demand, with reference to an increase of power when wanted.

The variations of construction as to diameter of wheel, size of buckets and number of streams applied, as before mentioned, also admits of adaptation to all conditions and requirements of service, either as to speed or power, in the most efficient way and at the smallest cost.

Transmission

Of the various methods of transmitting power from the water wheel to the generator, that of using belting is to be avoided whenever possible, on account of the added floor space, loss in transmission and cost of maintenance. The methods most commonly employed in modern hydro-electric plants, all of which practically embrace direct-connection of wheel and generator shaft, are described below:

FLEXIBLE COUPLING: This form of coupling is furnished in halves, one pressed on the water wheel shaft, and the other half bored and key-seated to gages of shaft and key, provided by the electrical manufacturer. The leather links connecting the two halves of coupling permit a slight displacement in the alignment of the shafts, such as that occasioned by the settling of the foundation, or a slight error in the original setting of the machine.

OVERHUNG CONSTRUCTION: In this instance the water wheel is mounted direct on the rotor shaft, which is extended for that purpose, and overhangs the bearing. This construction is very desirable where the type of generator admits, as the water wheel and generator are thus one integral machine, without possibility of its getting out of alignment. In particular, the engine-type generator is well adapted to this construction.

DOUBLE OVERHUNG CONSTRUCTION: This is usually employed with the engine-type generator of large capacity, in which case the armature is mounted on a bed-plate with a bearing on either side, and the shaft carries a wheel at each end, overhanging the bearing, with the rotor in the center. In such cases the water wheel builder furnishes a bed-plate, shaft and bearings common to both apparatus, and of dimensions to suitably carry the engine-type generator.

FLANGED SOLID COUPLINGS should be used only on small, self-contained units where mounted on a common bed-plate and assembled by the manufacturer, as they are not susceptible of adjustment for alignment. It is always advisable to drive exciters by independent wheels, and direct-connection can usually be obtained, the overhung construction being generally adopted.

Illustrations will be found on succeeding pages, showing the various mountings and transmission methods above described.

Construction

In every instance the highest grade material and best workmanship are expended on all constructions with a view of obtaining the most reliable and efficient apparatus. All shafting, screw-threads, bolts and other parts are made true to Brown & Sharpe—U. S. Standard gages. All materials—such as cast-iron, bronze, gun metal, cast-steel, forgings and steel plates—are rejected unless up to the most rigid specifications, to insure those qualities that contribute to perfect results. Wheels are assembled as far as possible in the shops, and all joints stamped where they match, to facilitate erection at power house.

Upon the water wheel as a prime mover depends the entire transmission plant, representing usually a large investment. It is, therefore, essential that all material be of the highest grade attainable. The requirements of a transmission plant are 24-hour service, three hundred and sixty-five days in the year,—and PELTON WHEELS are constructed with this object in view!

Comparison With Turbines

As has been before stated, the PELTON WHEEL is not adapted for extremely low heads, the construction of the turbine wheel enabling it to handle large quantities of water to better advantage under such conditions.

With clear water under medium falls it is conceded that the turbine is ordinarily reliable and some of the best types highly efficient. The construction of the average turbine, however, is necessarily such that the vanes and operating gates are subject to wear, which causes a rapid impairment of efficiency. This is especially true where the water carries sand and grit, as is so often the case. All mountainous regions, particularly those in tropical countries, such as Mexico, Central and South America, present conditions which offer the most serious objections to the turbine wheel. The streams furnishing power in such localities are subject to sudden freshets from excess of rainfall, and carry, at such times, grit and sand sufficient to destroy any turbine in a very short time.   They also carry roots, leaves and other trash that fill the vanes and choke

the wheels to such an extent as to often prevent their running until the obstructions are removed, involving a degree of unreliability that discredits the many advantages such a power ought always to afford.

The PELTON WHEEL, on the contrary, is constructed with a perfectly free discharge, and the buckets will not choke up by anything that may be thrown upon them, thus making it absolutely reliable.

The Quintex-nozzle wheel, described and illustrated on page 32, is of special form and construction, intended particularly to meet such conditions as are here mentioned.

To cover conditions under which the PELTON WHEEL is not applicable, this Company has designed the PELTON-FRANCIS turbine — a full description of which, together with its advantages and limitations, will be found on page 102 of catalog.   It will thus be noted that this Company is prepared to cover the entire hydraulic field regardless of head, speed or power, and to furnish such apparatus as will be best suited to the existing conditions.

Speed Control

Under average conditions of operation, a governor is not necessary, as, with a constant load, the speed of the wheel is absolutely uniform.   When slight and infrequent changes occur—such as are caused by hanging up stamps of a battery, for example— the wheel can be regulated by hand, by means of the main stop gate.

When the changes of load are sudden and severe, as is particularly the case when operating electric power plants, an automatic governor is essential.   Under these conditions the speed of the wheel is controlled by means of various devices operated by the governor, among which may be described the following:

The DEFLECTING NOZZLE is a cast-iron or steel nozzle provided with a ball and socket joint, permitting of its being raised or lowered, thus throwing the stream on or off the buckets; the power of the wheel is consequently increased or diminished to adapt to the change of load, and a constant speed is maintained.   A steel deflecting plate, which deflects the stream itself — the nozzle remaining stationary — is sometimes used to accomplish the same results, when the design will not admit of a deflecting nozzle.

The STREAM CUT-OFF is a spherical plate fitting tightly over the end of the nozzle-tip, which, by varying its position, changes the discharge area of the nozzle, and thus influences the power of the wheel.

The NEEDLE NOZZLE consists of a nozzle body in which is inserted a concentric tapered needle. A change of position of this needle produces a corresponding change of discharge area of the nozzle; the amount of water used is thus varied and the power of the wheel influenced proportionately.

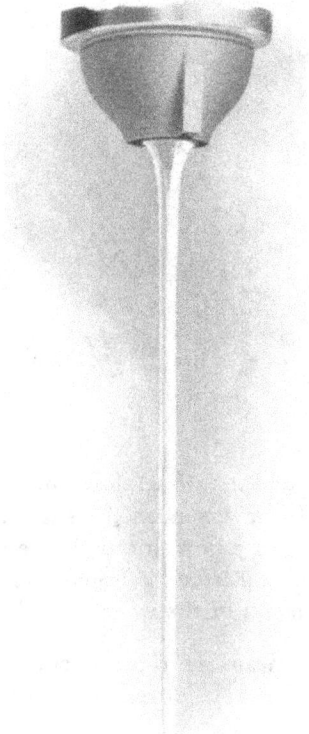

The NEEDLE AND DEFLECTING NOZZLE is a most valuable combination, consisting of a deflecting nozzle, with which is incorporated a needle nozzle with means for operating either the needle or deflecting nozzle simultaneously or separately. This accomplishes a two-fold object — accurate regulation and water economy without water ram. The deflecting nozzle, in itself, is a most sensitive means of regulation when actuated by an automatic governor but does not save water. On the other hand, the needle nozzle, while it is extremely economical in the use of water, is difficult to control quickly by means of the governor. The operation of the combination is as follows:

Assuming the full load to be on the water wheel and the nozzle in position of greatest efficiency, a decrease in load will cause the nozzle to be suddenly deflected by the automatic governor. Simultaneously, the needle portion of the nozzle will be actuated by hand, or by another automatic device, tending to *gradually* close the needle and decrease the flow. The governor then raises the nozzle to accommodate the decreased flow of water (and consequent decrease of power), and the nozzle is then brought back to the position of greatest efficiency, having, at the same time, controlled the speed within the required limits. Such a device is essential where water is valuable, and where economy is necessary to carry over the peak load. The needle portion need not necessarily be operated by an automatic device, but may be controlled by hand, and the same results obtained, although necessarily in a longer period of time. See illustration of this nozzle on page 92.

The conditions as to head, power and character of load determine which device is best adapted. These various mechanisms are actuated, through a proper system of rock-shafts and levers, by an automatic governor, description of which will be found on page 101.

Miscellaneous Notes

The power of wheels listed is in all cases based upon effective heads — no allowance being made for friction loss. In ordering wheels the maximum power desired should always be given; also full data regarding water supply. In this connection note carefully the information called for on page 27.

Different heads cannot be utilized in one pipe line or on the same wheel. Two or more wheels may be run on the same shaft under different heads, by proportioning their diameters to the respective heads involved.

Under high heads, where regulation is required, it is advisable to use the deflecting nozzle or the needle deflecting nozzle in combination, if water is scarce, which admits of

power variation without variation in pressure, thereby relieving the pipe line of shock due to water hammer.

The most economical results are obtained when running with the gate wide open and with a nozzle-tip of such size as to give the necessary power. When the water supply is diminished, the size of nozzle-tip should be reduced, or the needle partially closed, so as to maintain full head in the pipe line.

Where the wheel does not run in the right direction for the connecting machinery, it may be reversed on the shaft and the water brought around to the other side by a long radius elbow; or, with the wheel reversed, the water may be delivered to the top of the wheel from the same side, this latter expedient involving a loss of head equal to the diameter of the wheel.

By means of rope or sprocket chain attached to hand wheel of gate, it may be operated at any convenient point, even a considerable distance from the wheel.

Experimental Laboratory

Not content with the experience gained and possible improvement to be noted from the observation of each wheel installed, this Company is continually engaged in scientific experiment for the determination of the correct design and proportion of the various elements entering into the construction of the PELTON WHEEL. Perhaps the most interesting apparatus for this purpose is the "Scobiscopic" device patented by this Company.

As has been said, the efficiency of the PELTON WHEEL is due to the shape of the bucket, and, as efficiency is a most important factor in the development of every power proposition, it is the bucket and the action of the water upon it which engage the most attention.

Bucket Design

In designing a water wheel bucket and attempting to carry out a particular theory, it is of course necessary to lay out the bucket shape mathematically, and then, following out well-known hydraulic laws in regard to the flow of water on surfaces, plot the curves and paths which the stream should follow.

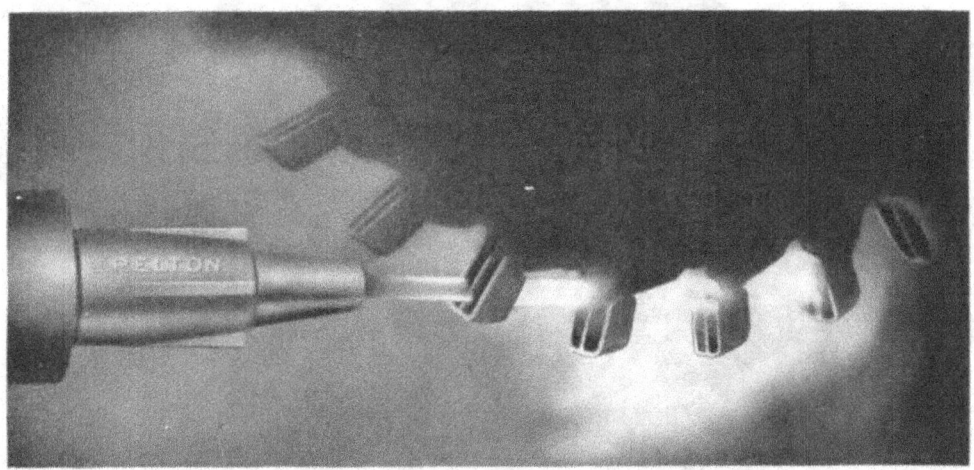

1002

Tangential Wheel with Pelton Buckets

Instantaneous photograph of tangential wheel fitted with PELTON buckets when running at high efficiency, showing the discharge from the sides of the buckets parallel with the entering jet; the photograph also shows clearly that the front of the bucket enters the stream without shock or disturbance of any kind and that all of the energy is removed from the water by this shape of bucket.

This plan, the only one heretofore known, has been productive of some excellent results, but the fallacy lies in the fact that most of the hydraulic laws above referred to are deduced from observations of water action against *stationary* surface, or, if moving, the action of the water has been so obscured by spray and the moving element that the observations have in some cases led to mistaken theories.

The problem is, then, to observe the action of the water on *moving* surfaces as though they are *standing still*.

Description of Apparatus

This is accomplished by applying the "Scobiscopic" device above mentioned to the observation of a water wheel in actual operation. The general principle involved is, that if a moving object be illuminated at fixed intervals by rays of light admitted at the moment the object passes the desired position, each portion of the object will be observed as though it were stationary. Thus is secured a clear view of what occurs in the buckets of a water wheel while it is operating under normal working conditions.

The " Scobiscopic " apparatus may be illustrated by the device shown below, which, while intended for exhibition purposes, embraces, on a small scale, the same mechanism as that employed in actual experimental work.

1003

This consists of a PELTON WHEEL direct connected to an electric generator—the water being supplied to the wheel through a centrifugal pump. The luminant is an electric search-light, which normally throws its direct rays on to the wheel, under which condition the wheel may be seen revolving and the water discharging from the buckets in apparent confusion. Interposed between the search-light and the wheel is a sheet iron disc with radial slots. The disc is revolved by means of gearing at a speed proportional to that of the wheel; when the search-light is brought into the upper position, the light is thrown on to the wheel *through* these revolving slots, and thus the wheel is illuminated at fixed intervals, as explained above, and one sees the wheel apparently standing still, and can observe the action of the water and trace the path of the jet from its entering the bucket to its reversal and clearance of the succeeding bucket, without interference.

Deductions

From a careful observation of this exhibit in operation, will be observed the following conditions, which are the main essentials to the attainment of a high bucket efficiency in an impulse water wheel :

First, that the front lip of the bucket in entering the stream of water produces absolutely no disturbance ; second, that the water velocity is taken up on the surface of the bucket in a line vertically under the shaft center, and when the bucket surface is in the most advantageous position ; third, that the water discharges from the sides of the buckets without interference with the succeeding buckets or the wheel center and with a minimum velocity of discharge ; fourth, that water is *not* carried around with the wheel, the discharge occurring as soon as the water has transferred its energy to the rotating wheel ; fifth, that no twist or disturbance occurs in the water jet ; hence the development of the greatest possible power from the moving water.

The bucket curvatures for obtaining these results as here mentioned have resulted in securing from operating wheels, bucket efficiencies in excess of ninety per cent, exclusive of journal friction and windage losses. In the actual construction of PELTON WATER WHEELS for high duty — such as power transmission plants — the buckets are always specially designed to secure best efficiency under the actual conditions of operation.

Engineering Practise

This Company is constantly experimenting with a view to improving on both the efficiency and mechanical workmanship of the PELTON WHEEL and its accessory apparatus. While the fundamental patents under which the PELTON WHEEL was first constructed made a vast step in the progress of hydraulic development, these original designs have been very materially improved upon. At the present time the standard type of PELTON bucket differs radically from the earlier forms, and in fact, is the result of a series of evolutions deduced from exhaustive experiment and calculation. A particular feature of PELTON engineering is the fact that practically every wheel is designed especially to meet the requirements of the work to be performed. For example, a type of PELTON bucket designed for a high head and small amount of power will differ, not only in size, but in general design, from that type which is recommended for use under a moderate head for a large amount of power. Experience has shown that, in order to obtain the highest useful effect from the water, the conditions surrounding each individual case must be carefully considered, and the apparatus constructed accordingly. Careful observation of these principles has resulted in the uniformly high efficiency and successful operation of the PELTON WHEEL wherever installed.

The succeeding pages are devoted to the description and illustration of the various standard and special wheels manufactured by this Company. Again is emphasized the fact that the greater part of PELTON apparatus is of special design, built for the particular work to be performed ; consequently it is essential that the fullest particulars of any proposed development be submitted, in order that proper adaptation of wheels may be made.

The Pelton Water Motor

The term "MOTOR," as used by this Company, is intended to mark a distinction between the smaller sizes of wheels, up to and including 24 inches diameter, and the STANDARD WHEELS, 3 feet and over, described on page 25.

Constructed entirely of cast-iron, the PELTON MOTOR is heavy and substantial, of compact form, and entirely self-contained. As sent out, it is ready for operation, needing no erection other than bolting to a solid foundation and making pipe connections. All PELTON MOTORS are provided with automatic, ring-oiling journal boxes, requiring a minimum of attention and insuring cleanliness. In addition are the centrifugal discs which effectually prevent water leaking along the shaft — a common fault with most water motors.

One particular feature of the PELTON MOTOR is its adaptability to varying capacities. With each motor are furnished two nozzle-tips of different diameters. By changing the size of stream on the wheel, a great variation in power may be obtained, and thus it is possible to have surplus power available for an emergency, while running normally with the least amount of water.

The PELTON MOTOR is identical in principle and operation with the standard PELTON WHEEL, and possesses all the advantages of the former as to efficiency and reliability. The Company claims that in detail of mechanical construction, design and finish, it has no equal in the market.

There are five sizes of standard PELTON MOTORS — varying from 6 inches to 24 inches in diameter — illustrations of which will be found on the opposite page. The standard sizes are designed as Type C motors, and their capacities under different pressures will be noted from tables on pages 28 to 31 inclusive. Motors of special sizes and capacities are made, to adapt them to any conditions of speed and power. Some applications of these motors to special purposes are illustrated on pages 22 and 23.

PELTON MOTORS are largely used for running dynamos in isolated electric plants; for this purpose, where the pressure admits, both water motor and dynamo are mounted on a cast-iron base-plate, and the shafts direct-connected by means of a coupling, thus occupying little floor space and eliminating belting. If the power is required for other use during the day, a pulley can be added and temporary belt connection made.

This motor is especially adapted for running from city mains, and, where water rates are at all reasonable, it will be found cheaper and much more convenient than any other means of producing power.

When water is limited, or paid for at meter rates, it is essential that the greatest amount of power be obtained from the least amount of water; in other words, that the motor be HIGHLY EFFICIENT. The PELTON MOTOR is guaranteed to fully realize in practical working all claims made, provided it is properly installed in accordance with instructions.

IT IS NEEDLESS TO SPECIFY THE GREAT VARIETY OF USES to which these motors are adapted, but the following list embraces some of the machinery to which they are applicable as power:

Dynamos for Electric Lights, Concentrators, Rock-breakers, Passenger and Freight Elevators, Printing Presses, Wine Presses, Power Pumps, Woodworking Machinery, General Machine Tools, Exhaust Fans or Blowers for ventilating mines, Ice Cream Freezers, Churns and various Dairy Machines, Dental Lathes, etc.

1304

Design of 6, 12 and 15-inch Type C Motor

Design of 18 and 24-inch Type C Motor

The Pelton Water Motor

Price List and Weights—Type C Motors

Size, Inches	Weight, Pounds	Pulley		Price
		Diameter, Inches	Face, Inches	
6	40	3	2	$30
12	160	6	4	60
15	265	7½	5	125
18	390	9	6	175
24	680	12	8	275

The above table covers standard motors for ordinary service up to a head of 230 feet—equivalent to a pressure of 100 pounds; all are provided with overhung pulleys.

For heads in excess of this a heavier motor is furnished, usually with a sole plate and outboard bearing, where the full power is required. *Special quotations* should be obtained to cover these conditions.

Prices include gate valve and nipple, driving pulley and two interchangeable nozzle-tips, to give variations in power. Unless otherwise specified, the standard size of driving pulley, as indicated above, will be furnished.

COMPARISON IN PRICE can only be made with reference to the work accomplished; hence, taking into consideration the capacity and efficiency of the PELTON MOTOR, it is proper to say that the prices above named are much lower than those of any other make.

ESSENTIAL DATA: The information called for on page 27, covering wheels, applies largely to motors as well; but to summarize in this particular case, especially in regard to motors operated from city mains, parties should give the following information: Head or pressure available in water main; size of main; length of pipe necessary from main to motor; number of bends or elbows (if any); amount of power required, and kind of machinery to be operated. Where possible, correspondents should state the dimensions and speed of pulley to which it is intended to belt from pulley on motor shaft.

SPEED OF MOTOR: The motor in all cases should be so geared as to allow it to run at speed indicated in tables under the pressure involved. This is accomplished by properly proportioning the diameters of driving and driven pulleys. The net pressure should be taken after deducting the loss of pressure due to friction in the connecting pipe from main to motor. This is best ascertained by placing a pressure gage directly back of the gate valve and noting the pressure registered when the wheel is running.

With a high pressure, when a small amount of power is required for running slow-moving machinery, it is often advisable to use a larger motor than is actually required for the power, on account of reducing the initial speed, thus rendering it easier to belt down to the speed of the driven machinery, and frequently saving the cost of counter-shafting and pulleys with their attendant loss of power.

By properly proportioning the nozzle-tip to the power, water will be used in direct ratio to the amount of power developed, regardless of the maximum capacity of the motor.

CHANGE OF TIPS: Two nozzle-tips are sent with each motor, the larger giving the maximum capacity. It is advisable to experiment with these, using the smaller first, and then the larger, if necessary to give the required power. To obtain the highest efficiency the motor should be run with the valve wide open. If this gives too much power, a smaller tip should be used. The change of tips is made by taking off the hand hole cover forming the name plate on side of case, and unscrewing the tip by means of a wrench. The pipe connections with gate should be made with a union.

SIZE OF PIPES: The diameter of SUPPLY PIPE to motor depends largely on the pressure available, length of pipe, and the amount of power required. Wherever possible, this information should be submitted, to admit of special calculation in this regard. As a guide, however, assuming an average pressure of, say, 60 pounds to the square inch, and a pipe about 100 feet in length, it may be stated that the supply pipes for various motors should not be less than the following: 6-inch motor, 2-inch pipe; 12-inch motor, 3-inch pipe; 15-inch motor, 3½-inch pipe; 18-inch motor, 4-inch pipe; 24-inch motor, 5-inch pipe.

There are sometimes restrictions as to size of taps which the water company will permit in its mains—in which case, if a single tap is not of sufficient size, it is advisable to use two or more small taps, and join them to a larger pipe a short distance from the main. Right-angle elbows should be avoided, as they materially reduce the pressure.

The DISCHARGE PIPE should be of ample diameter to carry the water away from the motor without backing up. It is generally advisable to set the motor on a small box or tank, allowing the water to discharge therein, then carrying it off, by means of piping, to a drain or other receptacle.

Special 24-inch Motor with Outboard Bearing—for High Duty
(Price on application)
A special circular, dealing exclusively with PELTON MOTORS, sent on request

21

Pelton Motor Direct-connected to Dynamo — Overhung Construction

Type of 18 and 24-inch Motor with Overhung Pulley

1008

Motor and Centrifugal Pump—Direct-connected

1009

Triplex Plunger Pump—Direct-geared to Pelton Water Motor

(Prices on application)

Standard Pelton Wheel Mounted on Wood Frame

1010

Standard Pelton Wheels

This Company manufactures a line of standard wheels from 3 feet to 6 feet in diameter, the capacities and speeds of which will be found on pages 28 to 31.

The standard PELTON WHEEL is usually mounted on a wood frame, as shown in cut on opposite page. Where heavy powers are involved, it is sometimes advisable to set the journals on masonry foundations and to enclose the wheel with a sheet steel cover housing.

These standard wheels are usually recommended for mining plants, sawmills, etc., where economy of floor space is no particular object, and where the self-contained, iron-mounted type of wheel, described on page 9, would be unnecessarily expensive.

As furnished, the standard wheel equipment includes the wheel with buckets, shaft, three journal boxes, two set collars, gate valve, cast-iron rigid nozzle and two interchangeable nozzle-tips to give the required power variation. Deflecting nozzles, stream cut-offs, needle nozzles, and governor connections are not a part of the standard wheel equipment, but may be readily supplied should the requirements demand.

It is usually advisable to build the wood frame and make the frame rods and bolts on the ground, thereby effecting a saving in freight to destination. If desired, the frame will be made at the works, the wheel erected thereon, then marked to match, taken down and packed for shipment, thereby enabling the work to be readily assembled on arrival. It will be found, however, that any millwright, or one reasonably familiar with such work, will have no trouble in constructing the frame from such detailed drawings as are furnished.

Connection between the PELTON gate and main pipe should be made by a flange joined to a taper or screwed pipe connection, matching the joint on main pipe. The flange and taper connection are not furnished, but can be supplied on receipt of data regarding character and size of main pipe.

The weights of the 3-foot and 4-foot wheels come within the limit for mule packing. The larger sizes of wheels for these requirements can be made in sections not exceeding 250 to 300 pounds.

Attention is again called to the fact that PELTON WHEELS are largely designed for the work to be performed; the Company would prefer in each individual case to decide as to whether the standard or special type of wheel is best adapted in order that satisfactory results may be obtained. To determine this, full data regarding conditions of operation are essential. Prices on all water wheel apparatus will be given on application.

See data required for estimates, page 27

Special Pelton Wheels

On the succeeding pages will be found illustrations and descriptions of the various wheel equipments comprising the PELTON System of Power. Each illustration is from a unit actually constructed, and from the diversity of arrangements will be noted the extreme flexibility of the system and its ready adaptation to all hydraulic conditions. No attempt has been made to establish list prices on any of this apparatus, which is built on order only, but detailed specifications and estimates will be furnished on receipt of necessary data.

See data required for estimates, page 27.

Terms and Other Information

Unless otherwise specified and previously arranged, terms to customers who have not established credit relations are always cash in current funds. Foreign customers should establish a credit with some bank in San Francisco or New York — depending on the office at which order is placed — with instructions to pay full amount of invoice on presentation of shipping documents.

While not responsible for either safe or prompt delivery after shipment in good condition, this Company will at all times cooperate with its patrons in having their claims equitably adjusted.

Preliminary plans on any proposed work submitted on receipt of sufficient data. Complete construction plans, embracing full details, are furnished with all orders.

This Company is especially well equipped for furnishing complete pipe lines with all connections, and requests the opportunity of figuring on same. There is a decided advantage in having one company responsible for both wheel and pipe line, as this insures accurate matching of the various connections.

An experience of many years in planning and executing enterprises of this character, with unvarying success, affords assurance that such information as may be given by this Company, regarding the best means of utilizing water powers, is entitled to the fullest confidence; also that all apparatus and material furnished will be adapted to the situation and requirements of the case, while involving such elements of security, economy and reliability as should attach to work of so important a character.

Where there is a substantial compliance with conditions given, as regards water supply and head, there can be no disappointment in results, as these can be calculated upon with all the certainty that applies to any problem in mathematics.

Attention is called to this Company's unusual facilities for supplying at the lowest ruling rates, riveted SHEET STEEL or LAP–WELD PIPE, together with RELIEF VALVES, WATER GATES, RECEIVERS and other connections; also SHAFTINGS, PULLEYS, COUPLINGS, JOURNAL BEARINGS, ROPE DRIVES, and other POWER–TRANSMITTING MACHINERY. Propositions submitted for entire power plants of any capacity and under any conditions of service.

Data Required for Estimates

PARTICULAR ATTENTION is directed to the following data, as its receipt is necessary for a proper understanding of any hydraulic proposition; a careful compliance therewith will save much time in correspondence:

FIRST: Amount of water available, either in miners' inches or the quantity in cubic feet or gallons *per minute*. If in miners' inches, state the head over the center of opening in measuring box, as the practise in regard to this varies in different localities. For data regarding measurement of water, see pages 40 to 44. If the quantity of water is not constant at all seasons, give the maximum and minimum flow. State also if it is desired to have the wheel utilize the maximum water supply.

SECOND: Head or *vertical* fall from ditch, flume or other source of supply to the point where wheel is to be located.

THIRD: Length of pipe necessary to obtain the given head. If already laid, give diameter; or, if of several sizes, give the length of each size. It is always advisable to furnish, if possible, a profile of the pipe line showing the pressure at different points, so as to enable an economic distribution of pipe gages proportionate to the pressures involved. If the line is of extreme length and under a heavy pressure, this data is very essential.

FOURTH: Horse-power required, maximum and minimum, and the character of machinery to be driven. Always give, if possible, dimensions and speed of pulley which is to be driven by pulley on water wheel shaft, together with size of keyway. If for driving an electric generator, state of what manufacture, its speed, kilo-watt capacity and size of shaft; and, if possible, furnish an outline drawing of the generator. State also whether the current is to be used for power transmission or lighting purposes, or both. If automatic regulation is desired, give the probable load variation and frequency of changes.

FIFTH: A sketch of the power house floor plan will facilitate an adaptation of the water wheel apparatus to local conditions. This should show the relative positions of the pipe line and the point where the water discharges from the wheel, together with an idea as to the surrounding country, in order that the tail-race may be properly designed.

SIXTH: When the size of flume is given, state the velocity of flow or the grade on which the flume is laid; also the depth of running water. If the actual carrying capacity is known, state this as well.

SEVENTH: *Do not* give as the water supply an amount that will fill a certain size of pipe, calculations based on such information being unreliable.

EIGHTH: Where reference is made to information contained in catalog, state the tables from which such information is obtained, the number of the illustration and the edition of the catalog.

HOME CORRESPONDENTS will please write the address plainly, giving post office with county, and state or territory.

FOREIGN CORRESPONDENTS should state full address, with name of post office, country, colony or province, and such other information as may be necessary to insure safe and prompt transmission.

Pelton Water Wheel Tables
Standard Sizes

The calculations for power in these tables are based upon the application of one stream to the wheel and on *effective* heads. In using these tables liberal allowance should be made to cover the friction loss in pipe, elbows, gates, etc. The light-face figures under those denoting the various heads give the equivalent pressure in pounds per square inch, and spouting velocity of water in feet per minute. The cubic measurement of water is also based on the flow per minute.

Head in Feet		Size of Wheels								
		6 Inch	12 Inch	15 Inch	18 Inch	24 Inch	3 Foot	4 Foot	5 Foot	6 Foot
20 8 Lbs. 2151.97	Horse-power	.05	.12	.20	.37	.66	1.50	2.64	4.18	6.00
	Cubic Feet	1.67	3.91	6.62	11.72	20.83	46.93	83.32	130.36	187.72
	Revolutions	629	315	252	210	157	105	79	63	52
30 13 Lbs. 2635.62	Horse-power	.10	.23	.38	.69	1.22	2.76	4.88	7.69	11.04
	Cubic Feet	2.05	4.79	8.11	14.36	25.51	57.44	102.04	159.66	229.76
	Revolutions	770	385	308	257	192	128	96	77	64
40 17 Lbs. 3043.39	Horse-power	.15	.35	.59	1.06	1.89	4.24	7.58	11.85	16.96
	Cubic Feet	2.37	5.53	9.37	16.59	29.46	66.36	107.84	184.36	265.44
	Revolutions	891	446	356	297	223	148	111	89	74
50 21 Lbs. 3402.61	Horse-power	.21	.49	.84	1.49	2.65	5.98	10.60	16.63	23.93
	Cubic Feet	2.64	6.18	10.47	18.54	32.93	74.17	131.72	206.13	296.70
	Revolutions	996	428	398	332	249	166	125	100	83
60 26 Lbs. 3727.37	Horse-power	.28	.65	1.10	1.96	3.48	7.84	13.94	21.77	31.36
	Cubic Feet	2.90	6.77	11.47	20.31	36.08	81.25	144.32	225.80	325.00
	Revolutions	1090	545	436	363	273	182	136	109	91
70 30 Lbs. 4026.00	Horse-power	.35	.82	1.39	2.47	4.39	9.88	17.58	27.51	39.52
	Cubic Feet	3.13	7.31	12.39	21.94	38.97	87.76	155.88	243.89	351.04
	Revolutions	1178	589	471	393	295	196	147	118	98
80 34 Lbs. 4303.99	Horse-power	.43	1.00	1.70	3.01	5.36	12.04	21.44	33.54	48.16
	Cubic Feet	3.35	7.82	13.25	23.46	41.66	93.84	166.64	260.73	375.36
	Revolutions	1258	629	503	419	315	210	157	126	105
90 39 Lbs. 4565.04	Horse-power	.51	1.20	2.03	3.60	6.39	14.40	25.59	40.04	57.60
	Cubic Feet	3.55	8.29	14.05	24.88	44.19	99.52	176.75	276.55	398.08
	Revolutions	1336	668	534	445	334	223	167	134	111
100 43 Lbs. 4812.00	Horse-power	.60	1.40	2.32	4.21	7.49	16.84	29.93	46.85	67.36
	Cubic Feet	3.74	8.74	14.81	26.22	46.58	104.88	186.32	291.51	419.52
	Revolutions	1408	704	563	469	352	235	176	141	117
110 48 Lbs. 5046.87	Horse-power	.69	1.62	2.74	4.86	8.64	19.44	34.58	54.11	77.76
	Cubic Feet	3.92	9.16	15.53	27.50	48.85	110.00	195.41	305.73	440.00
	Revolutions	1477	738	591	492	369	246	185	148	123
120 52 Lbs. 5271.30	Horse-power	.79	1.84	3.12	5.54	9.85	22.18	39.41	61.66	88.75
	Cubic Feet	4.10	9.57	16.21	28.72	51.02	114.91	204.10	319.33	459.64
	Revolutions	1543	771	617	514	386	257	193	154	129
130 56 Lbs. 5486.54	Horse-power	.89	2.08	3.53	6.25	11.11	25.02	44.46	69.53	100.08
	Cubic Feet	4.27	9.96	16.89	29.90	53.10	119.60	212.43	332.37	478.41
	Revolutions	1606	803	642	535	402	268	201	161	134
140 60 Lbs. 5693.65	Horse-power	.99	2.33	3.94	6.99	12.41	27.96	49.64	77.71	111.85
	Cubic Feet	4.43	10.34	17.53	31.03	55.11	124.12	220.44	344.92	496.48
	Revolutions	1667	833	667	556	417	278	208	167	139
150 65 Lbs. 5893.44	Horse-power	1.10	2.58	4.37	7.75	13.77	31.01	55.08	86.22	124.04
	Cubic Feet	4.55	10.70	18.14	32.11	57.04	128.43	228.19	357.02	513.90
	Revolutions	1725	862	690	575	431	288	216	172	144
160 69 Lbs. 6086.74	Horse-power	1.22	2.84	4.82	8.54	15.17	34.16	60.68	94.94	136.65
	Cubic Feet	4.73	11.05	18.74	33.17	58.92	132.68	235.68	368.73	530.75
	Revolutions	1783	891	713	594	446	297	223	178	149

Pelton Water Wheel Tables
Standard Sizes

The calculations for power in tables below are based upon the application of one stream to the wheel and on *effective* heads. In using these tables liberal allowance should be made to cover the friction loss in pipe, elbows, gates, etc. The light-face figures under those denoting the various heads give the equivalent pressure in pounds per square inch, and spouting velocity of water in feet per minute. The cubic measurement of water is also based on the flow per minute.

Head in Feet		Size of Wheels								
		6 Inch	12 Inch	15 Inch	18 Inch	24 Inch	3 Foot	4 Foot	5 Foot	6 Foot
170 74 Lbs. 6274.07	Horse-power Cubic Feet Revolutions	1.33 4.88 1835	3.11 11.39 917	5.28 19.31 734	9.35 34.19 612	16.61 60.73 459	37.42 136.77 306	66.46 242.93 229	103.99 380.08 183	149.68 547.08 153
180 78 Lbs. 6455.97	Horse-power Cubic Feet Revolutions	1.45 5.02 1885	3.39 11.72 942	5.75 19.87 754	10.19 35.18 628	18.10 62.49 471	40.77 140.74 314	72.41 249.97 236	113.30 391.10 188	163.08 562.96 157
190 82 Lbs. 6632.86	Horse-power Cubic Feet Revolutions	1.57 5.16 1937	3.68 12.04 968	6.24 20.41 775	11.05 36.14 646	19.63 64.20 484	44.21 144.59 323	78.53 256.82 242	122.87 401.81 194	176.86 578.38 161
200 87 Lbs. 6805.17	Horse-power Cubic Feet Revolutions	1.70 5.29 1987	3.97 12.36 998	6.74 20.94 795	11.93 37.08 662	21.20 65.87 497	47.75 148.35 331	84.81 263.49 248	132.70 412.25 199	191.00 593.40 165
210 91 Lbs. 6973.26	Horse-power Cubic Feet Revolutions	1.83 5.42 2037	4.28 12.66 1018	27.25 21.46 815	12.84 38.00 679	22.81 67.50 509	51.38 152.01 340	91.26 270.00 255	142.78 422.44 204	205.52 608.06 170
220 95 Lbs. 7137.35	Horse-power Cubic Feet Revolutions	1.96 5.55 2086	4.59 12.96 1043	7.77 21.96 834	13.77 38.89 695	24.46 69.08 522	55.09 155.59 348	97.85 276.35 261	153.10 432.38 209	220.36 622.36 174
230 100 Lbs. 7297.78	Horse-power Cubic Feet Revolutions	2.10 5.68 2133	4.90 13.25 1066	8.31 22.46 853	14.72 39.77 711	26.15 70.64 533	58.89 159.08 355	104.60 282.56 267	163.66 442.09 213	235.56 636.35 178
240 105 Lbs. 7454.70	Horse-power Cubic Feet Revolutions	2.24 5.80 2160	5.23 13.54 1080	8.86 22.93 864	15.69 40.62 720	27.87 72.16 540	62.77 162.50 360	111.50 288.64 270	174.45 451.60 216	251.10 650.03 180
250 108 Lbs. 7608.44	Horse-power Cubic Feet Revolutions	2.38 5.92 2224	5.56 13.82 1112	9.42 23.42 890	16.68 41.46 741	29.63 73.64 556	66.74 165.86 371	118.54 294.59 278	185.47 460.91 222	266.96 663.45 185
260 113 Lbs. 7759.10	Horse-power Cubic Feet Revolutions	2.52 6.04 2269	5.89 14.09 1134	10.05 23.88 907	17.69 42.28 756	31.43 75.10 567	70.78 169.14 378	125.72 300.43 284	196.71 470.04 227	283.15 676.59 189
270 118 Lbs. 7906.93	Horse-power Cubic Feet Revolutions	2.67 6.15 2313	6.24 14.36 1156	10.67 24.34 925	18.72 43.09 771	33.26 76.53 578	74.90 172.36 385	133.05 306.15 289	208.17 479.00 231	299.63 689.46 193
280 121 Lbs. 8052.01	Horse-power Cubic Feet Revolutions	2.82 6.26 2357	6.59 14.62 1178	11.16 24.79 943	19.77 43.88 786	35.12 77.94 589	79.11 175.53 393	140.51 311.77 295	219.84 487.79 236	316.44 702.12 196
290 126 Lbs. 8194.54	Horse-power Cubic Feet Revolutions	2.97 6.38 2398	6.94 14.88 1199	11.77 25.23 959	20.84 44.66 799	37.02 79.32 599	83.38 178.64 400	148.10 317.29 300	231.73 496.42 240	333.55 714.56 200
300 130 Lbs. 8334.62	Horse-power Cubic Feet Revolutions	3.13 6.48 2440	7.31 15.13 1220	12.38 25.66 976	21.93 45.42 813	38.95 80.67 610	87.73 181.69 407	155.83 322.71 305	243.82 504.91 244	350.94 726.76 203
310 134 Lbs. 8472.39	Horse-power Cubic Feet Revolutions	3.29 6.59 2481	7.68 15.39 1240	13.01 26.08 992	23.04 46.17 827	40.92 82.01 620	92.16 184.69 414	163.69 328.04 310	256.11 513.25 248	368.64 738.78 207

Pelton Water Wheel Tables

Standard Sizes

The calculations for power in tables below are based upon the application of one stream to the wheel and on *effective* heads. In using these tables liberal allowance should be made to cover the friction loss in pipe, elbows, gates, etc. The light-face figures under those denoting the various heads give the equivalent pressure in pounds per square inch, and spouting velocity of water in feet per minute. The cubic measurement of water is also based on the flow per minute.

Head in Feet		6 Inch	12 Inch	15 Inch	18 Inch	24 Inch	3 Foot	4 Foot	5 Foot	6 Foot
320 139 Lbs. 8607.94	Horse-power	3.45	8.05	13.64	24.16	42.91	96.65	171.68	268.60	386.62
	Cubic Feet	6.70	15.63	26.50	46.91	83.32	187.65	333.29	521.46	750.60
	Revolutions	2520	1260	1008	840	630	420	315	252	210
330 143 Lbs. 8741.43	Horse-power	3.61	8.43	14.29	25.30	44.94	101.22	179.72	281.29	404.89
	Cubic Feet	6.80	15.88	26.91	47.64	84.61	190.56	338.46	529.55	762.24
	Revolutions	2558	1279	1023	852	640	427	320	256	213
340 147 Lbs. 8872.89	Horse-power	3.78	8.82	14.94	26.46	47.00	105.86	188.02	294.18	423.44
	Cubic Feet	6.90	16.12	27.31	48.35	85.88	193.42	343.55	537.51	773.71
	Revolutions	2597	1298	1039	866	649	433	325	260	216
350 152 Lbs. 9002.43	Horse-power	3.94	9.21	15.61	27.64	49.09	110.56	196.38	307.25	442.27
	Cubic Feet	7.00	16.35	27.71	49.06	87.14	195.25	348.57	545.36	785.00
	Revolutions	2636	1318	1054	879	659	439	330	264	220
360 156 Lbs. 9130.14	Horse-power	4.10	9.61	16.28	28.83	51.21	115.34	204.86	320.52	461.36
	Cubic Feet	7.10	16.58	28.10	49.75	88.37	199.03	353.51	553.10	796.14
	Revolutions	2674	1337	1070	891	668	446	335	267	223
370 160 Lbs. 9256.02	Horse-power	4.29	10.01	16.97	30.04	53.36	120.18	213.45	333.97	480.72
	Cubic Feet	7.20	16.81	28.49	50.44	89.59	201.78	358.39	560.73	807.12
	Revolutions	2710	1355	1084	902	678	452	339	271	226
380 165 Lbs. 9380.32	Horse-power	4.46	10.42	17.66	31.27	55.54	125.08	222.16	347.60	500.33
	Cubic Feet	7.30	17.04	28.88	51.12	90.80	204.48	363.20	568.25	817.95
	Revolutions	2746	1373	1099	915	687	458	343	275	229
390 169 Lbs. 9502.93	Horse-power	4.64	10.83	18.36	32.51	57.75	130.05	231.00	361.41	520.20
	Cubic Feet	7.39	17.26	29.25	51.79	91.98	207.16	367.95	575.68	828.64
	Revolutions	2782	1391	1112	927	695	464	348	278	232
400 173 Lbs. 9624.00	Horse-power	4.82	11.25	19.07	33.77	59.98	135.08	239.94	375.40	540.35
	Cubic Feet	7.49	17.48	29.63	52.45	93.16	209.80	372.64	583.02	839.20
	Revolutions	2818	1409	1126	939	705	470	352	282	235
410 178 Lbs. 9743.57	Horse-power	5.00	11.68	19.79	35.04	62.24	140.18	248.99	389.57	560.75
	Cubic Feet	7.58	17.70	30.00	53.10	94.31	212.40	377.26	590.26	849.63
	Revolutions	2854	1427	1141	951	713	476	357	285	238
420 182 Lbs. 9861.66	Horse-power	5.19	12.11	20.52	36.33	64.54	145.34	258.16	403.91	581.39
	Cubic Feet	7.67	17.91	30.36	53.74	95.46	214.98	381.84	597.41	859.93
	Revolutions	2890	1445	1156	963	723	482	361	289	241
430 186 Lbs. 9978.35	Horse-power	5.37	12.54	21.26	37.64	66.86	150.57	267.44	418.42	602.28
	Cubic Feet	7.76	18.12	30.72	54.38	96.59	217.52	386.36	604.48	870.11
	Revolutions	2923	1461	1169	974	731	487	365	292	244
440 191 Lbs. 10093.74	Horse-power	5.56	12.98	22.01	38.96	69.20	155.85	276.82	433.11	623.40
	Cubic Feet	7.85	18.33	31.07	55.01	97.70	220.04	390.82	611.47	880.16
	Revolutions	2956	1478	1182	985	739	493	370	296	246
450 195 Lbs. 10207.79	Horse-power	5.75	13.43	22.76	40.29	71.57	161.19	286.31	447.95	644.78
	Cubic Feet	7.94	18.54	31.42	55.63	98.81	222.52	395.24	618.38	890.11
	Revolutions	2989	1494	1196	996	747	498	373	299	249
460 200 Lbs. 10320.58	Horse-power	5.95	13.88	23.53	41.65	73.97	166.60	295.91	462.97	666.40
	Cubic Feet	8.03	18.74	31.77	56.24	99.90	224.98	399.61	625.22	899.95
	Revolutions	3022	1511	1209	1007	755	504	378	302	252

Pelton Water Wheel Tables
Standard Sizes

The calculations for power in tables below are based upon the application of one stream to the wheel and on *effective* heads. In using these tables liberal allowance should be made to cover the friction loss in pipe, elbows, gates, etc. The light-face figures under those denoting the various heads give the equivalent pressure in pounds per square inch, and spouting velocity of water in feet per minute. The cubic measurement of water is also based on the flow per minute.

Head in Feet		Size of Wheels								
		6 Inch	12 Inch	15 Inch	18 Inch	24 Inch	3 Foot	4 Foot	5 Foot	6 Foot
470 204 Lbs. 10432.17	Horse-power	6.14	14.33	24.29	43.01	76.40	172.06	305.61	478.15	688.25
	Cubic Feet	8.12	18.95	32.11	56.85	100.98	227.42	403.93	631.98	909.68
	Revolutions	3055	1527	1222	1018	764	509	382	306	254
480 208 Lbs. 10542.56	Horse-power	6.34	14.79	25.07	44.39	78.85	177.58	315.42	493.49	710.33
	Cubic Feet	8.20	19.15	32.45	57.45	102.05	229.82	408.20	638.66	919.29
	Revolutions	3088	1544	1234	1029	772	515	386	309	257
490 212 Lbs. 10651.79	Horse-power	6.54	15.26	25.80	45.79	81.33	183.16	325.32	509.00	732.65
	Cubic Feet	8.29	19.35	32.98	58.05	103.10	232.20	412.43	645.28	928.83
	Revolutions	3119	1559	1247	1040	780	520	390	312	260
500 217 Lbs. 10759.96	Horse-power	6.74	15.73	26.66	47.20	83.83	188.80	335.34	524.66	755.20
	Cubic Feet	8.37	19.54	33.12	58.64	104.15	234.56	416.62	651.83	938.25
	Revolutions	3152	1576	1261	1051	788	525	394	315	263
520 226 Lbs. 10973.04	Horse-power	. . .	. . .	. . .	. . .	. . .	200.22	355.62	556.39	800.88
	Cubic Feet	. . .	. . .	. . .	. . .	. . .	239.21	424.87	664.74	956.84
	Revolutions	. . .	. . .	. . .	. . .	. . .	535	401	321	267
540 234 Lbs. 11182.07	Horse-power	. . .	. . .	. . .	. . .	. . .	211.88	376.33	588.80	847.52
	Cubic Feet	. . .	. . .	. . .	. . .	. . .	243.76	432.96	677.41	975.07
	Revolutions	. . .	. . .	. . .	. . .	. . .	545	409	327	272
560 243 Lbs. 11387.26	Horse-power	. . .	. . .	. . .	. . .	. . .	223.76	397.43	621.82	895.04
	Cubic Feet	. . .	. . .	. . .	. . .	. . .	248.24	440.91	689.84	992.96
	Revolutions	. . .	. . .	. . .	. . .	. . .	556	417	334	278
580 252 Lbs. 11588.83	Horse-power	. . .	. . .	. . .	. . .	. . .	235.86	418.92	655.43	943.44
	Cubic Feet	. . .	. . .	. . .	. . .	. . .	252.63	448.71	702.04	1010.54
	Revolutions	. . .	. . .	. . .	. . .	. . .	566	424	340	283
600 260 Lbs. 11786.94	Horse-power	. . .	. . .	. . .	. . .	. . .	248.16	440.77	689.63	992.65
	Cubic Feet	. . .	. . .	. . .	. . .	. . .	256.95	456.38	714.05	1027.80
	Revolutions	. . .	. . .	. . .	. . .	. . .	575	431	345	287
650 282 Lbs. 12268.24	Horse-power	. . .	. . .	. . .	. . .	. . .	279.82	497.01	777.62	1119.29
	Cubic Feet	. . .	. . .	. . .	. . .	. . .	267.44	475.02	743.21	1069.77
	Revolutions	. . .	. . .	. . .	. . .	. . .	599	449	359	299
700 304 Lbs. 12731.34	Horse-power	. . .	. . .	. . .	. . .	. . .	312.73	555.46	869.06	1250.92
	Cubic Feet	. . .	. . .	. . .	. . .	. . .	277.54	492.95	771.26	1110.16
	Revolutions	. . .	. . .	. . .	. . .	. . .	621	466	373	310
750 326 Lbs. 13178.19	Horse-power	. . .	. . .	. . .	. . .	. . .	346.83	616.03	963.82	1387.34
	Cubic Feet	. . .	. . .	. . .	. . .	. . .	287.28	510.25	798.33	1149.13
	Revolutions	. . .	. . .	. . .	. . .	. . .	643	482	386	321
800 348 Lbs. 13610.40	Horse-power	. . .	. . .	. . .	. . .	. . .	382.09	678.66	1061.81	1528.36
	Cubic Feet	. . .	. . .	. . .	. . .	. . .	296.70	526.99	824.51	1186.81
	Revolutions	. . .	. . .	. . .	. . .	. . .	664	498	398	332
900 391 Lbs. 14436.00	Horse-power	. . .	. . .	. . .	. . .	. . .	455.94	809.82	1267.02	1823.76
	Cubic Feet	. . .	. . .	. . .	. . .	. . .	314.70	558.96	874.53	1258.81
	Revolutions	. . .	. . .	. . .	. . .	. . .	705	529	423	352
1000 434 Lbs. 15216.89	Horse-power	. . .	. . .	. . .	. . .	. . .	534.01	948.48	1483.97	2136.04
	Cubic Feet	. . .	. . .	. . .	. . .	. . .	331.72	589.19	921.83	1326.91
	Revolutions	. . .	. . .	. . .	. . .	. . .	742	556	446	371

1011

Quintex-nozzle Pelton Wheel with Wood Frame

Quintex-nozzle Pelton Wheels

The QUINTEX-NOZZLE PELTON WHEEL shown above was designed some years ago to meet the demand for a wheel to handle large quantities of water under moderately low heads. The wheel itself is of the characteristic PELTON type, but, as will be noted, the nozzle is curved to the periphery of the wheel and has five rectangular openings. These openings are controlled by a slide gate which closes them successively—thus affording a wide range of capacity and adaptation to varying water supply without loss of efficiency.

It will be observed from tables on page 34 that these wheels can be operated under any head to which a turbine is applicable; while possessing a free discharge, they afford the same advantages of efficiency and reliability as the standard PELTON WHEEL.

The illustration is that of the WOOD FRAME CONSTRUCTION, the frame work being generally built on the ground from detailed drawings furnished, a saving in freight being thereby effected.

In localities where woodwork is difficult to obtain, or climatic conditions do not favor its use, is recommended the SEMI-MASONRY CONSTRUCTION, as shown by plate on opposite page. In this construction the wheel is mounted on a cast-iron sole plate, carrying the journals and cast-iron housings.

1012

Quintex-nozzle Wheel—Semi-masonry Mounted

This sole plate, together with the nozzle support, is imbedded in concrete, making a solid and substantial construction which is especially desirable where large amounts of power are involved.

THE QUINTEX-NOZZLE WHEEL is also furnished in the IRON-MOUNTED type, which consists of wheel completely enclosed in cast-iron upper and lower housings, carrying the nozzle and journals. The wheel is thus entirely self-contained, and provides the added advantage of being more compact, occupying less floor space. This is necessarily the more expensive construction and is used only in special cases.

It should be borne in mind that the various mountings of these wheels do not affect the capacity or efficiency of a given size, the circumstances surrounding each case determining which type is best suited. The tables on page 34 indicate the capacity and speed of the various sizes of these wheels under given heads.

This type of wheel will be found particularly adapted as a substitute for the average low-head turbine, as the absence of vanes and delicate gate mechanism enables it to handle water carrying grit or debris without clogging or undue wear which causes such a rapid deterioration in efficiency. Parties should in all cases give full particulars as to the operating conditions. Prices to cover the various mountings of these wheels on application.

Quintex-nozzle Pelton Wheel

Capacity and Speed

Head in Feet		Size of Wheels				
		18 Inch	24 Inch	36 Inch	42 Inch	48 Inch
10	Horse-power . .	.82	1.19	2.66	4.66	7.36
	Revolutions . .	141	106	71	60	53
15	Horse-power . .	1.51	2.18	4.89	8.71	13.5
	Revolutions . .	172	129	86	74	65
20	Horse-power . .	2.32	3.35	7.50	13.4	20.8
	Revolutions . .	199	149	100	85	75
25	Horse-power . .	3.24	4.68	10.5	18.7	29.1
	Revolutions . .	223	167	111	95	84
30	Horse-power . .	4.25	6.15	13.8	24.6	38.2
	Revolutions . .	244	183	122	103	92
35	Horse-power . .	5.31	7.76	17.4	30.9	48.1
	Revolutions . .	263	197	132	113	99
40	Horse-power . .	6.58	9.44	21.2	37.9	59.2
	Revolutions . .	282	212	141	121	105
45	Horse-power . .	7.81	11.3	25.4	45.2	70.2
	Revolutions . .	299	224	150	128	112
50	Horse-power . .	9.18	13.2	29.6	52.8	82.1
	Revolutions . .	315	236	158	135	118
55	Horse-power . .	10.5	15.2	34.4	60.9	94.5
	Revolutions . .	331	248	165	142	124
60	Horse-power . .	11.9	17.4	39.2	69.6	108.
	Revolutions . .	345	259	173	148	129
65	Horse-power . .	13.6	19.7	44.2	77.6	122.
	Revolutions . .	359	269	180	154	135
70	Horse-power . .	15.2	21.9	49.3	87.6	137.
	Revolutions . .	373	279	186	160	140
75	Horse-power . .	16.8	23.6	54.6	97.2	151.
	Revolutions . .	386	289	193	165	145
80	Horse-power . .	18.4	26.7	60.1	107.	166.
	Revolutions . .	399	299	199	171	149
85	Horse-power . .	20.3	29.2	65.9	117.	182.
	Revolutions . .	410	308	205	176	154
90	Horse-power . .	22.1	31.9	72.0	128.	199.
	Revolutions . .	423	317	211	181	158
100	Horse-power . .	25.9	37.4	84.1	149.	232.
	Revolutions . .	446	334	223	191	167

1013

Double Unit—Direct-connected

Exciter Units

The above illustrates a generator direct-connected to a double PELTON WHEEL unit. In this instance there were two wheels on one shaft, for the reason that the head was too low to admit of obtaining sufficient power on a single wheel, considering the high speed involved. In the case of exciters, it is always advisable to drive by an independent water wheel, rather than by belting from the generator shaft. By the latter arrangement any variation in speed of generator produces a corresponding change in exciter speed, consequently the excitation of the generator fields is varied, thus affecting the entire system. An independent driven exciter at all times delivers a constant current to the fields and thereby insures a steady output.

An ingenious method of driving exciters is that of placing the exciter between the water wheel and an induction motor, the shafts all being direct-connected. The advantage of this arrangement is that if the water wheel should fall off in power output, due to nozzle becoming plugged, or any other unforeseen cause, the induction motor, being in synchronism with the main generator, would take up the load and retain the speed at the normal. On the other hand, if there was a tendency for the water wheel to run away, the induction motor would act as a generator, passing its current into the main system, thus forcing the exciter to maintain its normal speed. It will be seen that more constant voltage is assured by this arrangement.

Hydraulic Tables

The tables on the pages following will be found to contain data regarding the power and measurement of water, which will be of interest to the practical man as well as the engineer. This Company, being the pioneer in the field of high-pressure power development, has had a wide experience which has enabled it to deduce many original and useful formulæ which may be relied upon in actual practise. These tables have been carefully revised, and are believed to be remarkably free from error.

Table for Calculating the Horse-power of Water on Pelton Wheels

The following table gives the horse-power of 1 cubic foot of water per minute under heads from 1 up to 2100 feet.

Heads in Feet	Horse-power	Heads in Feet	Horse-power
1	.0016098	430	.692214
20	.032196	440	.708312
30	.048294	450	.724410
40	.064392	460	.740508
50	.080490	470	.756606
60	.096588	480	.772704
70	.112686	490	.788802
80	.128784	500	.804900
90	.144892	520	.837096
100	.160980	540	.869292
110	.177078	560	.901488
120	.193176	580	.933684
130	.209274	600	.965880
140	.225372	650	1.046370
150	.241470	700	1.126860
160	.257568	750	1.207350
170	.273666	800	1.287840
180	.289764	850	1.368330
190	.305862	900	1.448820
200	.321960	950	1.529310
210	.338058	1000	1.609800
220	.354156	1050	1.690290
230	.370254	1100	1.770780
240	.386352	1150	1.851270
250	.402450	1200	1.931760
260	.418548	1250	2.012250
270	.434646	1300	2.092740
280	.450744	1350	2.173230
290	.466842	1400	2.253720
300	.482940	1450	2.334210
310	.499038	1500	2.414700
320	.515136	1550	2.495190
330	.531234	1600	2.575680
340	.547332	1650	2.656170
350	.563430	1700	2.736660
360	.579528	1750	2.817150
370	.595626	1800	2.897640
380	.611724	1850	2.978130
390	.627822	1900	3.058620
400	.643920	1950	3.139110
410	.660018	2000	3.219600
420	.676116	2100	3.380580

When the Exact Head is Found in Above Table

EXAMPLE: Have 100-foot head and 50 cubic feet of water per minute. How many horse-power?

By reference to the above table the horse-power of each cubic foot under 100-foot head will be found to be .16098. This amount multiplied by the number of cubic feet per minute, 50, will give 8.04 horse-power.

When Exact Head is Not Found in Table

Take the horse-power of 1 cubic foot per minute under 1 foot head, and multiply by the number of cubic feet available, and then by the number of feet head. The product will be the required horse-power.

NOTE.—The above tables are based upon an efficiency of 85 per cent.

Table of Pelton Nozzles

Table giving the approximate discharge in *cubic feet per minute*, and horse-power developed by different stream diameters under various heads, when used on PELTON WHEELS.

	Discharge and Horse-power					Constants for Different Diameters	
Head in Feet	Discharge from 1-inch Stream	Horse-power from 1-inch Stream	Head in Feet	Discharge from 1-inch Stream	Horse-power from 1-inch Stream	Diameter	Constant
20	11.72	.37	420	53.74	36.33	.1	.01
30	14.36	.69	430	54.38	37.64	.2	.04
40	16.59	1.06	440	55.01	38.96	.3	.09
50	18.54	1.49	450	55.63	40.29	.4	.16
60	20.31	1.96	460	56.24	41.65	.5	.25
70	21.94	2.47	470	56.85	43.01	.6	.36
80	23.46	3.01	480	57.45	44.39	.7	.49
90	24.88	3.60	490	58.05	45.79	.8	.64
100	26.22	4.21	500	58.64	47.20	.9	.81
110	27.50	4.86	520	59.80	50.05	1.0	1.
120	28.72	5.54	540	60.94	52.97	.1	1.21
130	29.90	6.25	560	62.06	55.94	.2	1.44
140	31.03	6.99	580	63.16	58.96	.3	1.69
150	32.11	7.75	600	64.24	62.04	.4	1.96
160	33.17	8.54	650	66.86	69.95	.5	2.25
170	34.19	9.35	700	69.38	78.18	.6	2.56
180	35.18	10.19	750	71.82	86.70	.7	2.89
190	36.14	11.05	800	74.17	95.52	.8	3.24
200	37.08	11.93	900	78.67	113.98	.9	3.61
210	38.00	12.84	1000	82.93	133.50	2.0	4.
220	38.89	13.77	1050	85.2	143.90	.1	4.41
230	39.77	14.72	1100	87.2	154.34	.2	4.84
240	40.62	15.69	1150	89.1	164.83	.3	5.29
250	41.46	16.68	1200	91.1	175.91	.4	5.76
260	42.28	17.69	1250	93.	187.02	.5	6.25
270	43.09	18.72	1300	94.8	198.32	.6	6.76
280	43.88	19.77	1350	96.6	209.81	.7	7.29
290	44.66	20.84	1400	98.4	221.69	.8	7.84
300	45.42	21.93	1450	100.1	233.30	.9	8.41
310	46.17	23.04	1500	101.8	245.74	3.0	9.
320	46.91	24.16	1550	103.5	258.12	.1	9.61
330	47.64	25.30	1600	105.2	270.89	.2	10.24
340	48.35	26.46	1650	106.8	283.55	.3	10.89
350	49.06	27.64	1700	108.4	296.58	.4	11.56
360	49.75	28.83	1750	110.	309.76	.5	12.25
370	50.44	30.04	1800	111.5	323.01	.6	12.96
380	51.12	31.27	1850	113.1	336.69	.7	13.69
390	51.79	32.51	1900	114.6	350.44	.8	14.44
400	52.45	33.77	1950	116.1	364.32	.9	15.21
410	53.10	35.04	2000	117.6	378.55	4.0	16.
						.1	16.81
						.2	17.64
						.3	18.49
						.4	19.36
						.5	20.25
						.6	21.16
						.7	22.09
						.8	23.04
						.9	24.01
						5.0	25.

Directions for Using the above Table

To find the discharge in cubic feet per minute of a given stream, ascertain from table the discharge of a 1-inch stream under the head involved, then multiply it by the constant opposite the given diameter; the result will be the quantity discharged in cubic feet per minute by a stream of given diameter.

EXAMPLE: To find the quantity of water discharged by a stream 2.7 inches diameter under an effective head of 490 feet: Opposite 490 feet head find 58.05 cubic feet—amount discharged by a 1-inch stream under this head. Opposite 2.7 inches diameter, find the constant 7.29; multiply 58.05 by 7.29, and obtain 423.18—the quantity in cubic feet per minute discharged by the given stream under head named. The power developed by this diameter stream would be 45.79 by 7.29 = 333.80 horse-power.

Given the head and quantity of water, to find size of stream: Divide the quantity of water in cubic feet per minute by the discharge of a 1-inch stream under head named; the result will be the constant of the required diameter. The diameter opposite the nearest constant (if it does not correspond exactly) will indicate the approximate stream diameter to discharge the given quantity.

In the same way substitute horse-power for water quantity and find corresponding results.

NOTE.—The above tables of horse-power are based upon an efficiency of 85 per cent.

Loss of Head in Pipe by Friction

The following table shows the loss of head by friction in each 100 feet in length of different diameters of pipe, when discharging the following quantities of water per minute:

Vel. in Feet per Second	Inside Diameter of Pipe in Inches											
	6		7		8		9		10		11	
	Loss of Head in Feet	Cubic Feet per Minute	Loss of Head in Feet	Cubic Feet per Minute	Loss of Head in Feet	Cubic Feet per Minute	Loss of Head in Feet	Cubic Feet per Minute	Loss of Head in Feet	Cubic Feet per Minute	Loss of Head in Feet	Cubic Feet per Minute
2.0	.39	23.5	.33	32.0	.30	41.9	.26	53.0	.23	65.4	.21	79.
2.2	.46	25.9	.40	35.3	.35	46.1	.31	58 3	.28	72.	.25	87.
2.4	.54	28.2	.46	38.5	.41	50.2	.36	63.6	.32	78.5	.29	95.
2.6	.63	30.6	.54	41.7	47	54.4	.42	68 9	.37	85.1	.34	103.
2.8	.72	32.9	.61	44.9	.54	58.6	.48	74.2	.43	91.6	.39	111.
3.0	.81	35.3	.69	48.1	.61	62.8	.54	79.5	.48	98.2	.44	119.
3.2	.91	37.7	.78	51.3	.68	67.0	.60	84.8	.54	105.	.49	127.
3.4	1.02	40.0	.87	54.5	.76	71.2	.68	90.1	.61	111.	.55	134.
3.6	1.13	42.4	.96	57.7	.84	75.4	.75	95.4	.67	118.	.61	142.
3.8	1.25	44.7	1.07	60.9	.93	79.6	.83	101.	.74	124.	.68	150.
4.0	1.37	47.1	1.17	64.1	1.02	83.7	.91	106.	.82	131.	.74	158.
4.2	1.49	49.5	1.28	67.3	1.12	87.9	.99	111.	.89	137.	.81	166.
4.4	1.62	51.8	1.39	70.5	1.22	92.1	1.08	116.	.97	144.	.88	174.
4.6	1.76	54 1	1.51	73.7	1.32	96.3	1.17	122.	1.05	150.	.96	182.
4.8	1.90	56.5	1.63	76.9	1.43	100.0	1.27	127.	1.14	157.	1.04	190.
5.0	2.05	58.9	1.76	80.2	1.54	105.	1.37	132.	1.23	163.	1.12	198.
5.2	2.21	61.2	1.89	83.3	1.65	109.	1.47	138.	1.32	170.	1.20	206.
5.4	2.37	63.6	2.03	86.6	1.77	113.	1.57	143.	1.41	177.	1.28	214.
5.6	2.53	65.9	2.17	89.8	1.89	117.	1.68	148.	1.51	183.	1.37	222.
5.8	2.70	68.3	2.31	93.0	2.01	121.	1.80	154.	1.61	190.	1.46	229.
6.0	2.87	70.7	2.46	96.2	2.15	125.	1.92	159.	1.71	196.	1.56	237.
7.0	3.81	82.4	3.26	112.0	2.85	146.	2.52	185.	2.28	229.	2.07	277.

Vel. in Feet per Second	Inside Diameter of Pipe in Inches											
	12		13		14		15		16		18	
	Loss of Head in Feet	Cubic Feet per Minute	Loss of Head in Feet	Cubic Feet per Minute	Loss of Head in Feet	Cubic Feet per Minute	Loss of Head in Feet	Cubic Feet per Minute	Loss of Head in Feet	Cubic Feet per Minute	Loss of Head in Feet	Cubic Feet per Minute
2.0	.198	94.	.183	110.	.169	128.	.158	147.	.147	167.	.132	212.
2.2	.234	103.	.216	121.	.200	141.	.187	162.	.175	184.	.156	233.
2.4	.273	113.	.252	133.	.234	154.	.218	176.	.205	201.	.182	254.
2.6	.315	122.	.290	144.	.270	167.	.252	191.	.236	218.	.210	275.
2.8	.360	132.	.332	156.	.308	179.	.288	206.	.270	234.	.240	297.
3.0	.407	141.	.375	166.	.349	192.	.325	221.	.306	251.	.271	318.
3.2	.457	151.	.422	177.	.392	205.	.366	235.	.343	268.	.305	339.
3.4	.510	160.	.471	188.	.438	218.	.408	250.	.383	284.	.339	360.
3.6	.566	169.	.522	199.	.485	231.	.452	265.	.425	301.	.377	382.
3.8	.624	179.	.576	210.	.535	243.	.499	280.	.468	318.	.416	403.
4.0	.685	188.	.632	221.	.587	256.	.548	294.	.513	335.	.456	424.
4.2	.749	198.	.691	232.	.641	269.	.598	309.	.561	352.	.499f	445.
4.4	.815	207.	.751	243.	.698	282.	.651	324.	.611	368.	.542	466.
4.6	.883	217.	.815	254.	.757	295.	.707	339.	.662	385.	.588	488.
4.8	.954	226.	.881	265.	.818	308.	.763	353.	.715	402.	.636	509.
5.0	1.028	235.	.949	276.	.881	321.	.822	368.	.770	419.	.685	530.
5.2	1.104	245.	1.020	287.	.947	333.	.883	383.	.828	435.	.736	551.
5.4	1.183	254.	1.092	298.	1.014	346.	.947	397.	.888	452.	.788	572.
5.6	1.26	264.	1.167	309.	1.083	359.	1.011	412.	.949	469.	.843	594.
5.8	1.34	273.	1.245	321.	1.155	372.	1.078	427.	1.011	486.	.899	615.
6.0	1.43	283.	1.325	332.	1.229	385.	1.148	442.	1.076	502.	.957	636.
7.0	1.91	330.	1.75	387.	1.630	449.	1.520	515.	1.430	586.	1.270	742.

EXAMPLE—Have 200 feet head and 600 feet of 11-inch pipe, carrying 119 cubic feet of water per minute. To find effective head: In upper right-hand column under 11-inch pipe, find 119 cubic feet; opposite this will be found the coefficient of friction for this amount of water, which is .44. Multiply this by the number of *hundred* feet of pipe, which is 6, and you will have 2.64 feet, which is the loss of head. Therefore, the effective head is 200 — 2.64 -197.36.

Loss of Head in Pipe by Friction

The following tables show the loss of head by friction in each 100 feet in length of different diameters of pipe, when discharging the following quantities of water per minute:

Inside Diameter of Pipe in Inches

Vel. in Feet per Second	20		22		24		26		28		30	
	Loss of Head in Feet	Cubic Feet per Minute	Loss of Head in Feet	Cubic Feet per Minute	Loss of Head in Feet	Cubic Feet per Minute	Loss of Head in Feet	Cubic Feet per Minute	Loss of Head in Feet	Cubic Feet per Minute	Loss of Head in Feet	Cubic Feet per Minute
2.0	.119	262.	.108	316.	.098	377.	.091	442.	.084	513.	.079	589.
2.2	.140	288.	.127	348.	.116	414.	.108	486.	.099	564.	.093	648.
2.4	.164	314.	.149	380.	.136	452.	.126	531.	.116	616.	.109	707.
2.6	.189	340.	.171	412.	.157	490.	.145	575.	.134	667.	.126	766.
2.8	.216	366.	.195	443.	.180	528.	.165	619.	.153	718.	.144	824.
3.0	.245	393.	.222	475.	.204	565.	.188	663.	.174	770.	.163	883.
3.2	.275	419.	.249	507.	.229	603.	.211	708.	.195	821.	.182	942.
3.4	.306	445.	.278	538.	.255	641.	.235	752.	.218	872.	.204	1001.
3.6	.339	471.	.308	570.	.283	678.	.261	796.	.242	923.	.226	1060.
3.8	.374	497.	.340	601.	.312	716.	.288	840.	.267	973.	.249	1119.
4.0	.410	523.	.373	633.	.342	754.	.315	885.	.293	1026.	.273	1178.
4.2	.449	550.	.408	665.	.374	791.	.345	929.	.320	1077.	.299	1237.
4.4	.488	576.	.444	697.	.407	829.	.375	973.	.348	1129.	.325	1296.
4.6	.529	602.	.482	728.	.441	867.	.407	1017.	.378	1180.	.353	1355.
4.8	.572	628.	.521	760.	.476	905.	.440	1062.	.409	1231.	.381	1414.
5.0	.617	654.	.561	792.	.513	942.	.474	1106.	.440	1283.	.411	1472.
5.2	.662	680.	.602	823.	.552	980.	.510	1150.	.473	1334.	.441	1531.
5.4	.710	707.	.645	855.	.591	1018.	.546	1194.	.507	1385.	.473	1590.
5.6	.758	733.	.690	887.	.632	1055.	.583	1239.	.542	1437.	.506	1649.
5.8	.809	759.	.735	918.	.674	1093.	.622	1283.	.578	1488.	.540	1708.
6.0	.861	785.	.782	950.	.717	1131.	.662	1327.	.615	1539.	.574	1767.
7.0	1.143	916.	1.040	1109.	.953	1319.	.879	1548.	.817	1796.	.762	2061.

Inside Diameter of Pipe in Inches

Vel. in Feet per Second	33		36		39		42		45		48	
	Loss of Head in Feet	Cubic Feet per Minute	Loss of Head in Feet	Cubic Feet per Minute	Loss of Head in Feet	Cubic Feet per Minute	Loss of Head in Feet	Cubic Feet per Minute	Loss of Head in Feet	Cubic Feet per Minute	Loss of Head in Feet	Cubic Feet per Minute
2.0	.073	712.	.066	848.	.061	995.	.057	1155.	.053	1325.	.050	1508.
2.2	.085	785.	.078	933.	.072	1094.	.067	1270.	.063	1456.	.059	1658.
2.4	.100	855.	.091	1018.	.084	1194.	.079	1385.	.073	1590.	.069	1809.
2.6	.115	927.	.104	1100.	.097	1294.	.090	1500.	.084	1721.	.079	1960.
2.8	.131	1000.	.119	1188.	.111	1394.	.103	1617.	.096	1855.	.090	2110.
3.0	.148	1070.	.135	1273.	.125	1442.	.117	1730.	.109	1987.	.102	2260.
3.2	.167	1140.	.152	1367.	.141	1591.	.131	1845.	.122	2120.	.115	2410.
3.4	.186	1210.	.169	1442.	.157	1690.	.146	1961.	.136	2250.	.128	2560.
3.6	.206	1282.	.188	1527.	.174	1790.	.162	2079.	.151	2382.	.142	2715.
3.8	.226	1355.	.207	1612.	.191	1891.	.178	2190.	.166	2515.	.156	2865.
4.0	.248	1425.	.228	1697.	.210	1990.	.195	2310.	.182	2650.	.171	3016.
4.2	.270	1495.	.249	1782.	.229	2091.	.213	2422.	.198	2780.	.186	3165.
4.4	.295	1568.	.271	1866.	.250	2190.	.232	2540.	.216	2910.	.203	3318.
4.6	.321	1640.	.294	1951.	.271	2290.	.252	2658.	.235	3045.	.220	3470.
4.8	.346	1710.	.318	2036.	.293	2389.	.270	2770.	.254	3180.	.238	3619.
5.0	.374	1780.	.342	2121.	.316	2490.	.294	2885.	.273	3310.	.256	3770.
5.2	.403	1852.	.368	2206.	.342	2590.	.317	3000.	.296	3442.	.278	3920.
5.4	.430	1922.	.394	2291.	.364	2689.	.338	3115.	.315	3578.	.295	4071.
5.6	.453	1995.	.421	2376.	.393	2790.	.374	3230.	.340	3710.	.319	4222.
5.8	.495	2065.	.450	2460.	.419	2886.	.389	3348.	.363	3840.	.340	4373.
6.0	.520	2140.	.479	2545.	.441	2986.	.408	3461.	.382	3970.	.358	4524.
7.0	.693	2495.	.636	2968.	.586	3484.	.545	4030.	.509	4638.	.476	5277.

The following formula, deduced by Wm. Cox, gives practically the same results as the above table and will be found useful in many instances. $F = \frac{L}{10000 D}(4 V^2 + 5 V - 2)$. Where F = friction head, L = length of pipe in feet; D = diameter of pipe in inches; V = velocity in feet per second.

Measurement of Water

The information heretofore published on this subject, so essential to the development of any hydraulic proposition, has been of such a technical character as to be of little use to anyone but an engineer; hence, the attempt in this catalog to explain in a practical way the various methods of water measurement.

The most accurate is the WEIR DAM method described at length on pages 41 and 42. Any proposition of importance should be measured by this means, which is recognized as the most accurate. THE MINERS' INCH, as applied to water power calculations, has been eliminated from the tables in this catalog as being obsolete and inaccurate, from an engineering standpoint; but, for the information of those who desire to employ this means of measurement, it will be found illustrated and described on page 44. A rough approximation of a water quantity is sometimes desired, in which case the method known as MEASUREMENT BY CROSS-SECTION AND VELOCITY is used. This may be described as follows:

Select a stretch on a stream or ditch which will afford as straight and uniform a course as possible, avoiding pools or obstructions to the normal flow. If the water is at any point carried in a flume, it is better to measure at this point. Lay off a distance of, say 100 feet; measure the width of flowing water at about six different places in this distance, and obtain the average width; likewise at these same points measure the depth of water at three or four places across the stream, and obtain the average depth. Next, drop a float in the water, noting the number of seconds it takes to travel the given distance. From this can be calculated the velocity of the water in feet per second. The cubic quantity is the product obtained by multiplying the average width in feet by the average depth in feet by the velocity, which (if in feet per second) will give the flow of the stream in cubic feet per second. From the figures so obtained it is advisable to deduct about twenty-five per cent, as the surface velocity of water is in excess of the actual average velocity.

40

1014

Illustration of Weir Dam

Explanation of Weir Dam Measurement

Place a board across the stream at some point which will allow a pond to form above. The board should have a notch cut in it with both side edges and the bottom sharply beveled toward the intake, as shown in the above cut. The bottom of the notch, which is called the "crest" of the weir, should be perfectly level and the sides vertical.

In the pond back of the weir, at a distance not less than the length of the notch, drive a stake near the bank, with its top precisely level with the crest.

By means of a rule, or a graduated stake as shown, measure the depth of water over the top of stake, making allowance for capillary attraction of the water against the sides of the weir. For extreme accuracy this depth may be measured to thousandths of a foot by means of a "hook gage," familiar to all engineers. Having ascertained the depth of water over the stake, refer to table on page 43, from which may be calculated the amount of water flowing over the weir.

There are certain proportions which must be observed in the dimensions of this notch. Its length, or width, should be between four and eight times the depth of water flowing over the crest of the weir. The pond back of the weir should be at least fifty per cent wider than the notch and of sufficient width and depth that the velocity of flow or approach be not over 1 foot per second. In order to obtain these results it is advisable to experiment to some extent. A suppositious case may better serve to explain this procedure:

41

Permanent Weir Dam in Tail-race—Showing Hook Gage in Center

EXAMPLE: First, roughly gage the stream to be measured by the cross-section and velocity method described on page 40. Suppose the width is found to be 4½ feet, the depth 1½ feet, and the average velocity 3 feet per second. The stream is then carrying approximately 1215 cubic feet per minute.

Next it is necessary to ascertain the size of weir which will flow approximately this amount of water, having in mind that the length of notch must be from four to eight times its depth. This is best determined by the tables on page 43. Try first a depth of, say, 8 inches, from which it will be found that every inch of length will deliver 9.05 cubic feet per minute. Dividing 1215 (the required cubic quantity) by 9.05 it is found that it will require a weir 134 inches in width, which is 16.7 times the depth and therefore much too great a proportion. It is therefore obvious that the weir must be deeper. Trying again a depth of 15 inches in the same manner, note that it will require a width of about 52 inches, which is 3.4 times the depth and therefore too small a proportion. By another trial it is found that a depth of 12 inches and a width of 73 inches, a proportion of practically 1 to 6, and within the limits mentioned above, will discharge the desired amount and is therefore the approximate size of weir required.

Take a board of sufficient length to reach well across the stream and cut the notch 73 inches long and about 16 inches deep, so that the depth may be increased over the calculated amount, if necessary. The board should be made wide enough to admit of the vertical distance from the bottom of the notch to the level of water on the downstream side being at least twice the depth of water flowing over the weir.

If having proceeded this far, the width of the pond should not be 50 per cent wider than the length of the notch, or if the velocity of flow should be in excess of 1 foot per second, the pond should be enlarged or deepened to obtain the required result. It is of course essential that there be no leaks around or under the weir, in order that all the water may pass through the notch. Canvas or sacking laid under the water against the dam will be found effective in this connection.

With the weir so constructed, measure the depth of water over the stake, and by use of the tables as indicated, ascertain the actual quantity flowing over the weir.

Table for Weir Measurement

Giving cubic feet of water per minute, that will flow over a weir 1 inch long and from $\frac{1}{8}$ to 20$\frac{7}{8}$ inches deep.

Depth Inches	$\frac{1}{8}$	$\frac{1}{4}$	$\frac{3}{8}$	$\frac{1}{2}$	$\frac{5}{8}$	$\frac{3}{4}$	$\frac{7}{8}$
0	.00	.01	.05	.09	.14	.19	.26	.32
1	.40	.47	.55	.64	.73	.82	.92	1.02
2	1.13	1.23	1.35	1.46	1.58	1.70	1.82	1.95
3	2.07	2.21	2.34	2.48	2.61	2.76	2.90	3.05
4	3.20	3.35	3.50	3.66	3.81	3.97	4.14	4.30
5	4.47	4.64	4.81	4.98	5.15	5.33	5.51	5.69
6	5.87	6.06	6.25	6.44	6.62	6.82	7.01	7.21
7	7.40	7.60	7.80	8.01	8.21	8.42	8.63	8.83
8	9.05	9.26	9.47	9.69	9.91	10.13	10.35	10.57
9	10.80	11.02	11.25	11.48	11.71	11.94	12.17	12.41
10	12.64	12.88	13.12	13.36	13.60	13.85	14.09	14.34
11	14.59	14.84	15.09	15.34	15.59	15.85	16.11	16.36
12	16.62	16.88	17.15	17.41	17.67	17.94	18.21	18.47
13	18.74	19.01	19.29	19.56	19.84	20.11	20.39	20.67
14	20.95	21.23	21.51	21.80	22.08	22.37	22.65	22.94
15	23.23	23.52	23.82	24.11	24.40	24.70	25.00	25.30
16	25.60	25.90	26.20	26.50	26.80	27.11	27.42	27.72
17	28.03	28.34	28.65	28.97	29.28	29.59	29.91	30.22
18	30.54	30.86	31.18	31.50	31.82	32.15	32.47	32.80
19	33.12	33.45	33.78	34.11	34.44	34.77	35.10	35.44
20	35.77	36.11	36.45	36.78	37.12	37.46	37.80	38.15

Example Showing the Application of the Above Table

Suppose the weir to be 72 inches long, and the depth of water over the stake to be 11$\frac{5}{8}$ inches. Follow down the left-hand column of the figures in the table until you come to 11 inches. Then run across the table on a line with the 11, until under $\frac{5}{8}$ on top line, you will find 15.85. This multiplied by 72, the length of weir, gives 1141.2, the number of cubic feet of water passing per minute.

NOTE.—The above table will give results sufficiently close for all practical purposes, but if extreme accuracy is essential, the following formula should be used, in connection with measurements obtained from the method described on previous pages:

$$Q = 3.33 \, (L - .2H) \, H^{\frac{3}{2}}$$

In the above, L = length of weir in FEET; H = head or depth of flow in FEET over weir, as measured on the stake; Q = cubic feet of water per SECOND.

43

THE PELTON WATER WHEEL COMPANY

1015

Miners' Inch Measuring Box

Explanation of Miners' Inch Measurement

The term "miners' inch" is of California origin and not known or used elsewhere, it being a method of measurement adopted by the various ditch companies in disposing of water to their customers. This measurement is gradually becoming obsolete, and its use should be discouraged as, for the purpose of accurate calculation, it must always be reduced to a cubic quantity. Further, the term is somewhat indefinite, for the reason that the water companies do not always use the same head over the center of the opening; as a result, the inch varies from 1.36 to 1.73 cubic feet per minute in different localities.

The measuring box is easily constructed, however, and this method may be sometimes employed to good advantage in approximating a water quantity. The most common measurement is an aperture 2 inches high and whatever length is required to pass the water through a plank 1¼ inches thick, as shown in plate above. The lower edge of the aperture should be 2 inches above the bottom of the measuring box, and the plank 5 inches high above the aperture, thus making a 6-inch head above the *center* of the stream. Each *square* inch in this opening represents a miners' inch; thus, the slide pulled out one-half inch will represent one square inch of opening, which will discharge one miners' inch of water. It is generally customary to mark off one inch divisions, ascribing to each the value of *two* miners' inches.

The miners' inch above described is equal to 1½ cubic feet of water per minute, which is the legal inch as established by the California legislature. The element of time is not considered in the calculation of the miners' inch; that is to say, one miners' inch means a continuous flow of that quantity of water.

Data Regarding Flumes and Ditches

The carrying capacity of a flume or ditch depends on its size, slope, and the nature of the surface over which the water flows. It is impossible to accurately calculate the flow without knowing all the conditions, but the tables on page 46 will give approximate results for known dimensions, assuming average conditions.

The best results are obtained when the depth of water is one-half of its width, and the tables are based on this proportion. Ditches are constructed with sloping sides, usually at an angle of about 60 degrees from the vertical. For measuring ditches, take the width at one-half the depth of water.

Table I is for earth ditches of uniform size, free from weeds, etc.

Table II is for flumes of unplaned timber, very smooth brick-work, or smooth cemented ditches.

Table III is for flumes of planed timber, carefully fitted so that there are no rough joints or butts.

In general, the velocity of water should not be greater than 7 to 8 feet per second in wooden flumes, nor more than 3 feet in earth ditches, and these velocities should be exceeded only by an experienced engineer. It is usually advisable to keep the velocities down to about two-thirds of the above values.

Flow in Flumes and Ditches

Velocity in feet per second, and quantity in cubic feet per minute. For various sizes and slopes. *Width and depth in feet.*

TABLE I

Earth Ditches of Uniform Section in Good Condition

Slope per Rod Inches		1 x ½ Feet	1½ x ¾ Feet	2 x 1 Feet	3 x 1½ Feet	4 x 2 Feet	5 x 2½ Feet	6 x 3 Feet	7 x 3½ Feet	8 x 4 Feet	9 x 4½ Feet	10 x 5 Feet
⅛	Velocity	0.46	0.64	0.82	1.1	1.4	1.6	1.9	2.1	2.3	2.5	2.7
	Cubic Feet	13	43	98	300	670	1200	2050	3080	4400	6050	8100
¼	Velocity	0.65	0.91	1.2	1.6	2.0	2.3	2.7	3.0	3.3	3.6	3.9
	Cubic Feet	19	61	140	430	960	1720	2900	4400	6300	8700	11700
½	Velocity	0.93	1.3	1.6	2.2	2.8	3.3	3.8	4.2	4.7	..	..
	Cubic Feet	28	87	190	590	1340	2480	4100	6150	9200	..	..
¾	Velocity	1.1	1.6	2.0	2.7	3.4	4.0	4.6	..	..	..	..
	Cubic Feet	33	108	240	730	1630	3000	4950	..	..	..	..

TABLE II

Rough Timber, Smooth Brick-work or Cement

Slope per Rod Inches		1 x ½ Feet	1½ x ¾ Feet	2 x 1 Feet	3 x 1½ Feet	4 x 2 Feet	5 x 2½ Feet	6 x 3 Feet	7 x 3½ Feet	8 x 4 Feet	9 x 4½ Feet	10 x 5 Feet
⅛	Velocity	1.2	1.6	1.9	2.6	3.2	3.7	4.2	4.6	5.0	5.4	5.8
	Cubic Feet	36	110	230	700	1530	2780	4500	6700	9600	13000	17400
¼	Velocity	1.7	2.2	2.7	3.7	4.5	5.2	5.9	6.5	7.1	7.7	..
	Cubic Feet	51	148	324	1000	2160	3900	6350	9500	13600	18700	..
½	Velocity	2.4	3.1	3.9	5.2	6.4	7.4	..	..	..	..	..
	Cubic Feet	72	208	468	1400	3060	5550	..	..	..	..	..
¾	Velocity	2.3	3.8	4.8	6.4	7.8	..	..	..	..	..	..
	Cubic Feet	87	256	575	1720	3740	..	..	..	..	..	..
1	Velocity	3.3	4.4	5.5	7.3	..	..	..	..	..	..	..
	Cubic Feet	99	296	660	1970	..	..	..	..	..	..	..

TABLE III

Well Built Flumes of Smooth Planed Timber

Slope per Rod Inches		1 x ½ Feet	1½ x ¾ Feet	2 x 1 Feet	3 x 1½ Feet	4 x 2 Feet	5 x 2½ Feet	6 x 3 Feet	7 x 3½ Feet	8 x 4 Feet	9 x 4½ Feet	10 x 5 Feet
⅛	Velocity	1.7	2.3	2.7	3.6	4.4	5.0	5.6	6.2	6.8	7.3	7.7
	Cubic Feet	51	155	324	970	2100	3750	6000	9100	13000	17700	23000
¼	Velocity	2.4	3.2	3.9	5.1	6.2	7.1	8.0	..	..	..	..
	Cubic Feet	72	216	468	1370	2980	5300	8600	..	..	..	..
½	Velocity	3.4	4.5	5.5	7.2	..	..	..	..	..	..	..
	Cubic Feet	102	300	660	1940	..	..	..	..	..	..	..
¾	Velocity	4.1	5.5	6.7	..	..	..	..	..	..	..	..
	Cubic Feet	122	370	805	..	..	..	..	..	..	..	..
1	Velocity	4.8	6.4	7.7	..	..	..	..	..	..	..	..
	Cubic Feet	143	430	925	..	..	..	..	..	..	..	..

46

Practical Illustration of Diverting Dam, Intake and Head Gate for Flume

Table of Riveted Hydraulic Pipe

Showing price and weight with safe head for various sizes of double riveted.
(Revised)

Diameter of Pipe in Inches	Thickness of Material, U. S. Standard Gage	Equivalent Thickness in Inches	Head in Feet that Pipe will Safely Stand	Weight per Lineal Foot in Pounds	Price per Foot	Diameter of Pipe in Inches	Thickness of Material, U. S. Standard Gage	Equivalent Thickness in Inches	Head in Feet that Pipe will Safely Stand	Weight per Lineal Foot in Pounds	Price per Foot
3	18	.05	810	2.25	$0.20	18	12	.109	295	25.25	$1.90
4	18	.05	607	3.00	.25	18	11	.125	337	29.00	2.10
4	16	.062	760	3.75	.35	18	10	.14	378	32.50	2.40
						18	8	.171	460	40.00	3.00
5	18	.05	485	3.75	.30						
5	16	.062	605	4.50	.45	20	16	.062	151	16.00	1.26
5	14	.078	757	5.75	.50	20	14	.078	189	19.75	1.54
						20	12	.109	265	27.50	2.10
6	18	.05	405	4.25	.44	20	11	.125	304	31.50	2.25
6	16	.062	505	5.25	.50	20	10	.14	340	35.00	2.50
6	14	.078	630	6.50	.56	20	8	.171	415	45.50	3.40
7	18	.05	346	4.75	.50	22	16	.062	138	17.75	1.40
7	16	.062	433	6.00	.56	22	14	.078	172	22.00	1.70
7	14	.078	540	7.50	.63	22	12	.109	240	30.50	2.25
8	16	.062	378	7.00	.65	22	11	.125	276	34.50	2.40
8	14	.078	472	8.75	.75	22	10	.14	309	39.00	2.80
8	12	.109	660	12.00	.94	22	8	.171	376	50.00	3.75
9	16	.062	336	7.50	.69	24	14	.078	158	23.75	1.80
9	14	.078	420	9.25	.88	24	12	.109	220	32.00	2.35
9	12	.109	587	12.75	1.06	24	11	.125	253	37.50	2.70
10	16	.062	307	8.25	.72	24	10	.14	283	42.00	2.95
10	14	.078	378	10.25	.82	24	8	.171	346	50.00	3.50
10	12	.109	530	14.25	1.14	24	6	.20	405	59.00	4.30
10	11	.125	607	16.25	1.25						
10	10	.14	680	18.25	1.50	26	14	.078	145	25.50	2.00
						26	12	.109	203	35.50	2.59
11	16	.062	275	9.00	.75	26	11	.125	233	39.50	2.87
11	14	.078	344	11.00	.94	26	10	.14	261	44.25	3.10
11	12	.109	480	15.25	1.25	26	8	.171	319	54.00	3.85
11	11	.125	553	17.50	1.44	26	6	.20	373	64.00	4.75
11	10	.14	617	19.50	1.62						
12	16	.062	252	10.00	.82	28	14	.078	135	27.25	2.12
12	14	.078	316	12.25	1.00	28	12	.109	188	38.00	2.75
12	12	.109	442	17.00	1.38	28	11	.125	216	42.25	3.00
12	11	.125	506	19.50	1.50	28	10	.14	242	47.50	3.20
12	10	.14	567	21.75	1.69	28	8	.171	295	58.00	4.15
13	16	.062	233	10.50	.90	28	6	.20	346	69.00	5.00
13	14	.078	291	13.00	1.12						
13	12	.109	407	18.00	1.50	30	12	.109	176	39.50	2.90
13	11	.125	467	20.50	1.65	30	11	.125	202	45.00	3.15
13	10	.14	522	23.00	1.80	30	10	.14	226	50.50	3.50
						30	8	.171	276	61.75	4.30
14	16	.062	216	11.25	.98	30	6	.20	323	73.00	5.25
14	14	.078	271	14.00	1.17	30	1/4	.25	404	90.00	6.50
14	12	.109	378	19.50	1.57						
14	11	.125	433	22.25	1.72	36	11	.125	168	54.00	3.80
14	10	.14	485	25.00	1.95	36	10	.14	189	60.50	4.30
15	16	.062	202	11.75	.96	36	3/16	.187	252	81.00	5.75
15	14	.078	252	14.75	1.28	36	1/4	.25	337	109.00	7.60
15	12	.109	352	20.50	1.75	36	5/16	.312	420	135.00	9.50
15	11	.125	405	23.25	1.95						
15	10	.14	453	26.00	2.10	40	10	.14	170	67.50	4.75
16	16	.062	190	13.00	1.05	40	3/16	.187	226	90.00	6.40
16	14	.078	237	16.00	1.20	40	1/4	.25	303	120.00	8.40
16	12	.109	332	22.25	1.70	40	5/16	.312	378	150.00	10.50
16	11	.125	379	24.50	1.85	40	3/8	.375	455	180.00	12.00
16	10	.14	425	28.50	2.00						
						42	10	.14	162	71.00	5.05
18	16	.062	168	14.75	1.20	42	3/16	.187	216	94.50	7.00
18	14	.078	210	18.50	1.40	42	1/4	.25	289	126.00	9.50
						42	5/16	.312	360	158.00	12.00
						42	3/8	.375	435	190.00	15.00

NOTE.—These prices are high, to cover market fluctuatons in raw material, and are only intended for approximate purposes. Net prices on application.

Sheet Steel Riveted Pipe

In no other part of the world have the various systems of fluming, ditching and piping of water been developed to such an extent as on the Pacific Coast, and nowhere else are the advantages of these methods of conveying water so well understood. It is probably quite within bounds to say that there are anywhere from six to seven thousand miles of canals and flumes on the Pacific Coast, and perhaps quite as large an amount of piping, the latter of sizes running from 4 to 60 inches in diameter, and carrying water in some instances under pressure equivalent to 1700 feet head.

The question of water conduit is everywhere one of so much importance and is so intimately connected with the utilization of power by the PELTON WHEEL, that special consideration is given to it in this connection. The use of sheet steel pipe for power and other hydraulic purposes is strictly of California origin, and though so extensively adopted in the Pacific Coast States, it is only within recent years that its advantages have come to be understood and appreciated in other parts of the country.

The general impression among engineers who have not made this subject a special study has been that heavy cast-iron pipe or lap-weld tubing was necessary to withstand any considerable pressure, or for any degree of permanency. This prejudice has been largely overcome by the results obtained in actual practise extending over a period of many years, and it has been clearly demonstrated that riveted pipe, if properly made and designed for the purpose intended, is fully equal to any other form of construction, besides being very much cheaper and lighter.

Next to the proper adaptation of wheel to conditions of service, the most important matter in a water power plant is the pipe line and its various connections. If not well proportioned with reference to diameter, thickness, bends and connections, the efficiency will be greatly impaired and much trouble is likely to result from its insecurity.

Riveted pipe, as ordinarily constructed, is made from sheet steel or iron plates of the requisite thickness to withstand the pressure, which are rolled up into the required diameter. The sheets are secured by a double row of rivets on the longitudinal or straight seams, and a single row of rivets on the circular seams. It can be made up into independent sections of any length required, the average for steamer or rail shipment being about 25 feet; when for transportation by mule-back, the sections are generally about 7 feet long.

Each section of pipe, when constructed, is completely immersed in a bath of hot asphaltum composition which permeates the pores of the metal, coating the pipe inside and out with a hard, smooth surface of asphaltum, which effectually protects it from corrosion and increases the carrying capacity of the pipe, on account of the smooth and even surface presented to the water flow. The sections of pipe may be joined in different ways, depending on the local conditions and pressure involved. These various connections are illustrated and described on pages 50 and 51.

Referring to illustrations on page 50, Figure 1 shows the SLIP JOINT, which has one end of a section of pipe slightly tapered; this is securely driven into the larger end of the corresponding section, and forms a perfectly water-tight joint. This style of joint, if properly made, is safe for heads up to about 300 feet.

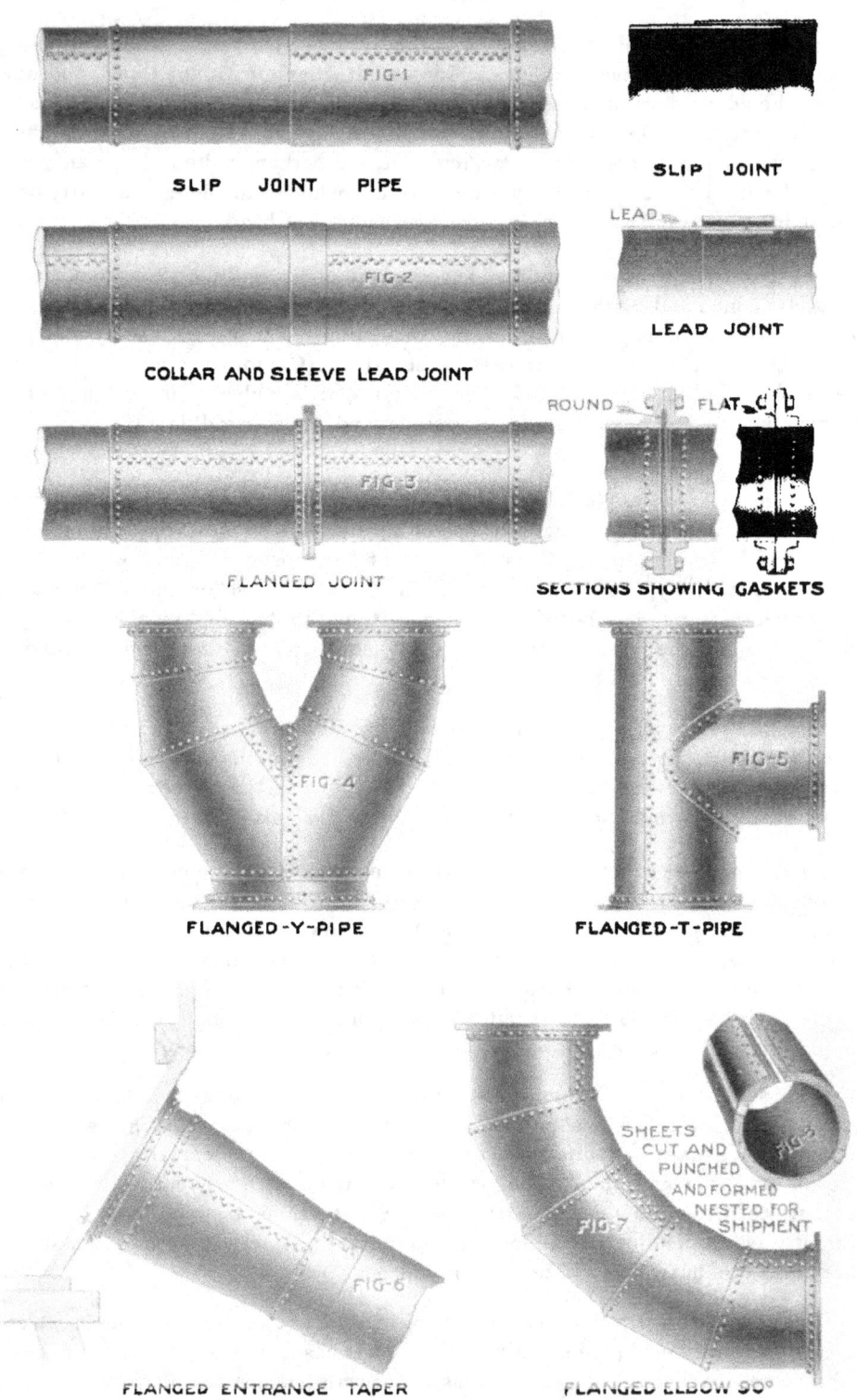

SLIP JOINT PIPE

COLLAR AND SLEEVE LEAD JOINT

FLANGED JOINT

SLIP JOINT

LEAD JOINT

SECTIONS SHOWING GASKETS

FLANGED-Y-PIPE

FLANGED-T-PIPE

FLANGED ENTRANCE TAPER
BOLTED TO PRESSURE BOX

FLANGED ELBOW 90°

Various Pipe Connections

Sheet Steel Riveted Pipe—*Continued*

The COLLAR AND SLEEVE LEAD JOINT is shown in Figure 2. In this joint the ends of the pipe are of the same size, and butt together. A sheet iron sleeve is riveted inside of one section of the pipe, one-half projecting beyond the joint, and a collar is placed in a corresponding position on the outside, leaving a space of about three-eighths of an inch between the collar and the pipe. This space is filled in with hot lead, which is securely caulked. The joint will stand a head of 700 feet, and has the advantage of allowing a slight adjustment of the angle — thus accommodating to any slight unevenness of the ground.

The illustration of the FLANGED JOINT (Figure 3) will be self-explanatory. A rubber gasket is placed between the faces of the flanges, which are drawn up by means of bolts. The gaskets may be round or flat, depending on certain conditions. The flanges may be made of cast-iron, rolled angle iron welded, or of pressed steel. These latter, which are pressed out of a flat steel plate by means of a die, are light and strong and cannot be broken. They are highly recommended for use under extreme pressure. There is practically no limit to the head under which flanged joints can be used, provided that they are properly proportioned.

Figures 4, 5 and 7 show the possibilities of sheet steel riveted pipe in making various connections. While flanged joints are illustrated, any style of joint is applicable.

Figure 6 illustrates one method of connecting the upper end of pipe line with the reservoir or flume. If the reservoir is of masonry, the flanges may be dispensed with.

Figure 8 shows a bundle of pipe cut, punched and formed, and nested for shipment. A pipe line is sometimes required in a locality where it is difficult to transport even short sections; in this case the sheets are cut to the required dimensions, punched for rivets, and formed to the proper diameter; several sheets are then nested together and clamped, thus occupying less space and making a bundle easily handled—a method of packing especially adapted for mule transportation. The sheets can then be riveted upon the ground to form the completed pipe, by means of tools furnished for that purpose.

In shipping pipe by steamer it is usually taken by cubic measurement (forty cubic feet to the ton) rather than by actual weight, and it is sometimes desirable to make a pipe line of several sizes, so that the smaller can be telescoped into the larger. In this way a very considerable saving in freight is effected.

When dealing with large, heavy pipe, it is often advisable to join the different sections by means of riveting them together in the field—thus forming a continuous riveted line. All sizes from 18-inch up admit of such connection.

This Company has large pipe works thoroughly equipped with the most modern machinery, affording every facility for turning out high-grade work. Estimates and specifications will be submitted covering pipe lines of any length, diameter or thickness. To this end there should be furnished a profile, showing the grade of the line and the angles involved, as also full data regarding water available and power required. Concise information on these points will insure the greatest economy in material and proper adaptation to conditions. Data and estimates also furnished regarding pipe lines, hydraulic giants and accessories for Placer mining.

All pipe orders are executed under the immediate supervision of the Company's engineers; workmanship is carefully inspected, and all angles checked up to templet; thus every pipe system *designed* and manufactured by the PELTON WATER WHEEL COMPANY carries a guarantee as to strength and durability, and complete fitness for the purpose intended.

Pelton Power Plant in Java

Note complete pipe line in background

5000 Horse-power Pelton Wheel

The above wheel, 9 feet 10 inches in diameter, is capable of developing 5000 horse-power
at 225 r. p. m., when operating under 865 feet effective head

018

Single Iron-mounted Unit—with Geared Gate Valve

One of four PELTON WHEEL units for direct-connection to electric generator. Head,
925 feet ; wheel, 84 inches diameter; 1500 horse-power capacity, at 333 r. p. m.

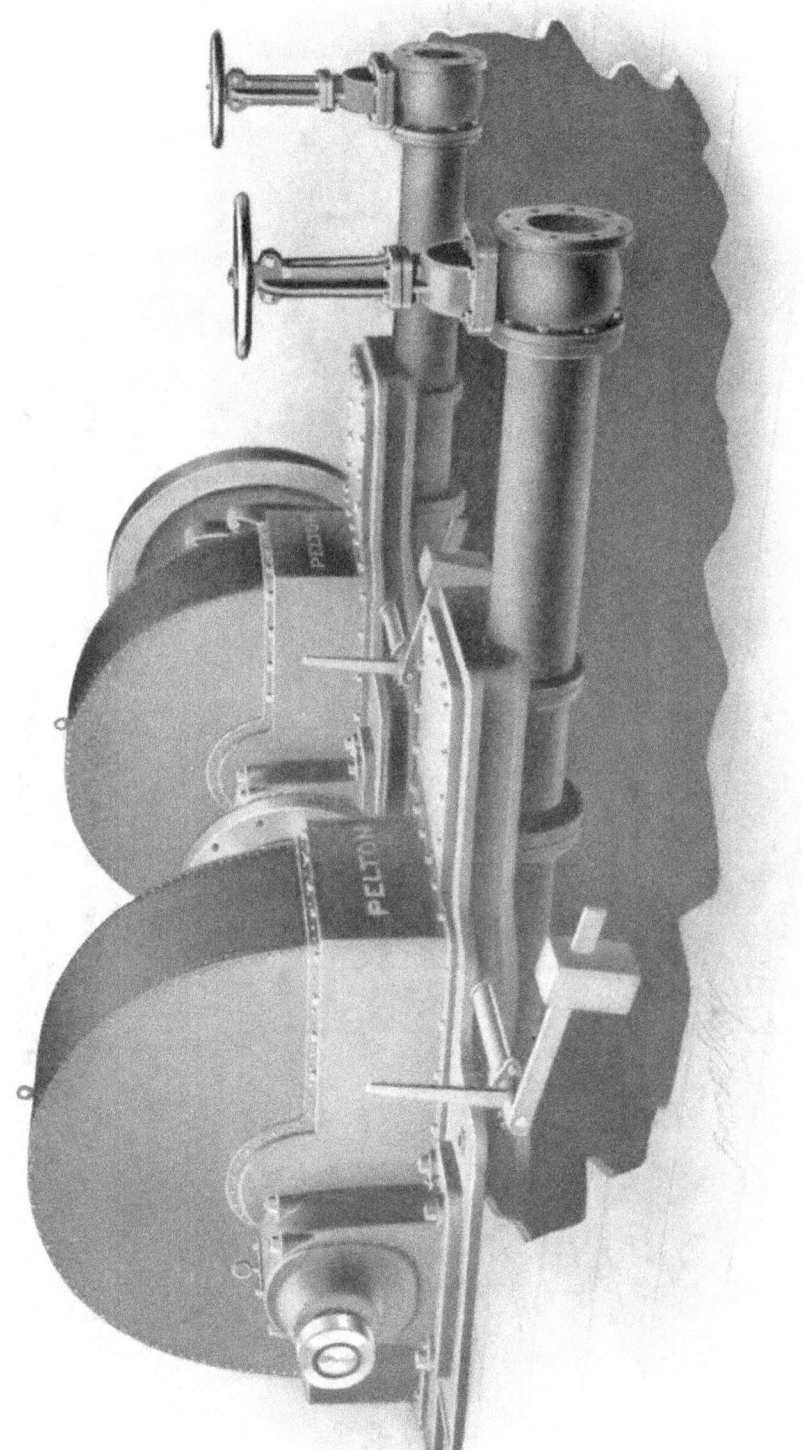

1019

Two Single Units for Direct-connection by means of Flexible Leather Link Couplings

58-inch diameter PELTON WHEELS.   Head, 590 feet ; each 600 horse-power capacity, at 400 r. p. m.   Typical example
of semi-masonry construction

Telluride Power Company—Single Iron-mounted Unit for Direct-connection to Generator

Showing needle nozzle mechanism below floor line and hydraulic-operated gate valve

Head, 900 feet, developing 3000 horse-power, at 225 r. p.m.

21

6

Telluride Power Company—Double Unit with Hydraulic-operated Gate Valves and By-passes

Head, 1800 feet; capacity, 4000 horse-power
Direct-connected to 2400 k.-w. 300 r. p. m. generator

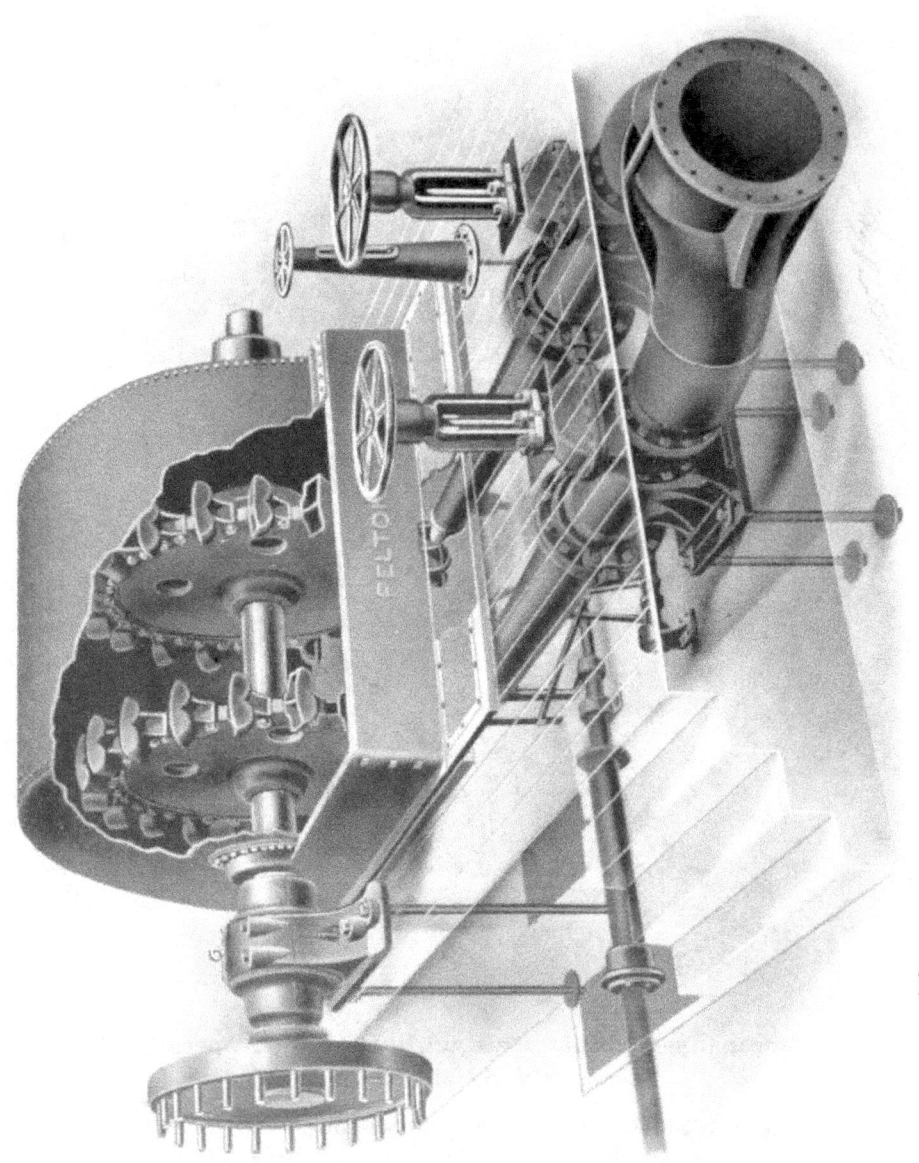

1020

Telluride Power Company—Double Unit for Direct-connection to Generator

2000 horse-power at 500 feet head.   Example of semi-masonry construction

The Telluride Power Company

The double unit illustrated on the opposite page is of 2000 horse-power capacity, running at 200 r. p. m., under a head of 500 feet, and is one of three units supplying power and light, by means of electric transmission, to a large group of mines in Colorado, which carry on a great variety of operations.

The wheels are each of cast-steel, and are fitted with steel buckets secured to the center by means of turned steel bolts. The wheels are pressed on a 10-inch diameter shaft, which is carried in " generator type " ring-oiling bearings. Only one-half of the coupling is shown, the other half being pressed on the shaft of the generator, which is of 1200 horse-power normal capacity. The coupling is of the flexible leather link type, each half being fitted with pins, as shown in the illustration, which are connected by means of leather links. This construction insures easy running, even though the alignment may not be true.

It will be seen that the wheels extend into the concrete foundation, and that the pipe, gate valves and nozzles are all carried in suitable channels in the concrete. The greatest possible rigidity results from this construction, as also, a freedom from vibration and noise, even in the largest sizes of wheels, thus making it an ideal arrangement, especially where the wheels are located within the power house, as is considered the best practise in modern transmission plants.

The shaft in this wheel, where it extends through the iron housing, is fitted with PELTON patented centrifugal discs and pockets—a device which successfully prevents leakage of water at the bearings and along the shaft, and at the same time eliminates the friction involved by the use of a stuffing-box.

The lower housing is constructed of cast-iron and carried well within the concrete foundation, and in front of this is a cast-iron floor plate, by the removal of which the nozzle, governor mechanism, and wheel pit may be reached for inspection.

The nozzles are of the ball and socket deflecting type, controlled by a governor, and one of the nozzles in addition is supplied with a stream cut-off which is operated from the floor-stand shown in front of the wheels. By means of this the station operator can regulate the flow of water to correspond with the maximum load at any hour of the day, and the governor will then, through deflecting nozzles, maintain the speed constant between this maximum and zero load.

The Telluride Power Company controls and operates a chain of hydro-electric plants in Colorado and Utah, having installed some 15,000 horse-power in PELTON WHEELS. These operate under varying conditions, ranging from 300 to 1725 feet in head and 400 to 5000 horse-power in unit capacity.

Illustrations on pages 56 and 57 describe two of the latest PELTON installations of this company.

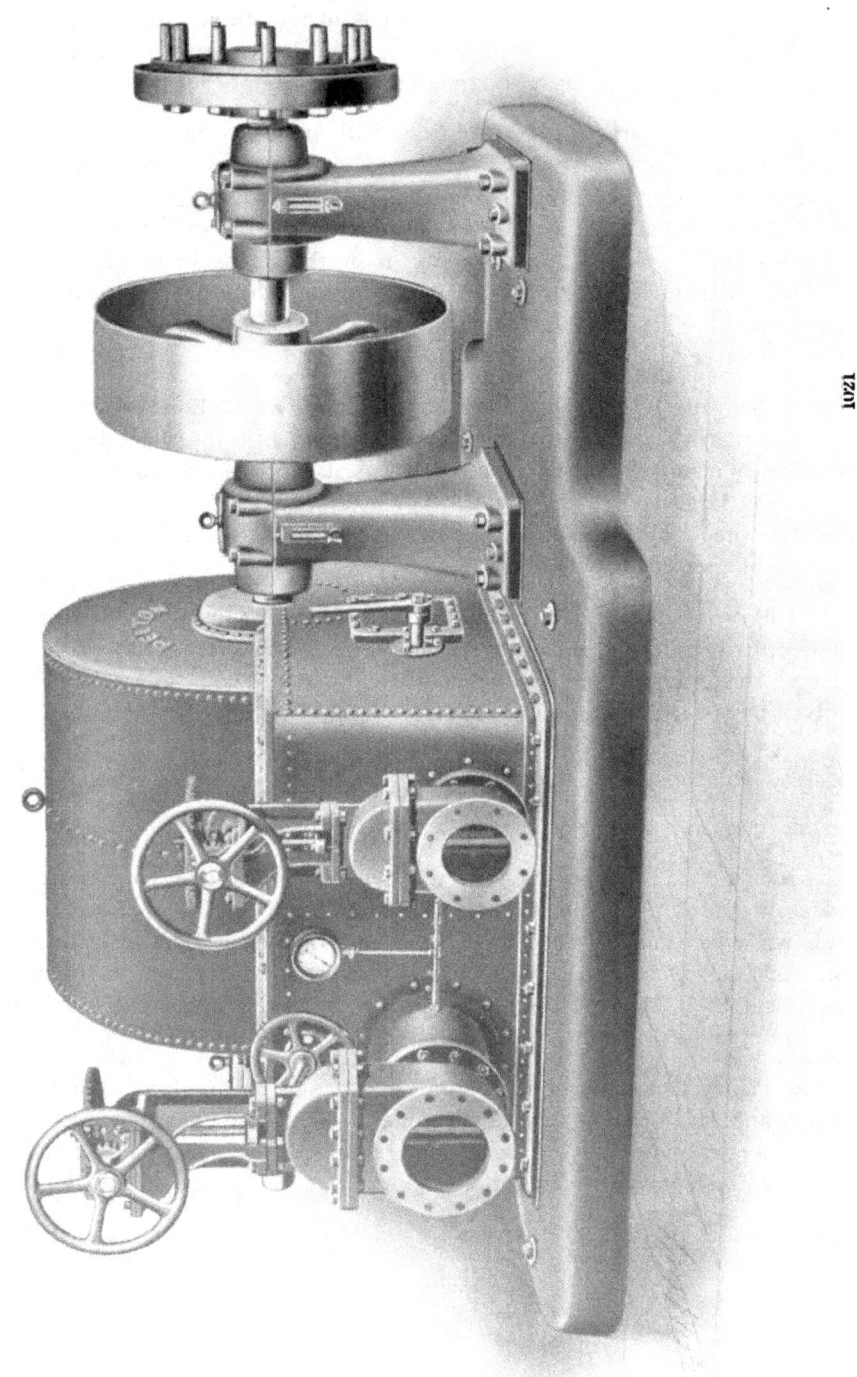

1021

Double Wheel Unit for Utilizing Two Different Water Heads

NOTE.—The wheels are of different diameters to adapt to heads of 450 and 700 feet respectively. The unit is normally for direct-connection, but at low water an auxiliary steam engine is belted to the pulley shown

13

Semi-masonry Double Unit

Head, 250 feet; 1000 horse-power, at 450 r. p. m.   Arranged for driving an electric generator at both ends of shaft by means of flexible leather link couplings

Double Unit—Semi-masonry Construction

For direct-connection to generator.   Head, 1150 feet ; 3000 horse-power, at 300 r. p. m.

Triple Unit—Semi-masonry Construction

Head, 142 feet ; 1500 horse-power, at 150 r. p. m.

1023

Air Compressors and Pelton Wheels

The above illustration is that of a PELTON WHEEL, mounted direct on the shaft of an Ingersoll-Rand (formerly the Ingersoll-Sergeant Drill Co.) compressor, and is one of three units furnished the Niagara Falls Power Company.

The wheel is approximately 7 feet in diameter, to admit of driving the compressor at its proper speed, under an effective head of 127 feet, and has a maximum capacity of 100 horse-power.

In order to provide fly-wheel effect, the cast-iron wheel center is of extra heavy construction, weighing in excess of 5000 pounds, and is made in halves to facilitate mounting on compressor shaft. Where transportation is difficult, the center may be constructed in several sections, so as to bring the weight of each piece within reasonable limits for handling.

The advantages of such direct connections are: Economy of power, as friction of belt or rope drive is eliminated, a minimum expense for maintenance, and saving of first cost of apparatus.

This application of the PELTON WHEEL is made with equal facility to all forms of compressors, as well as to blowers and many other classes of machinery. In such cases the wheels may be constructed of any diameter, ranging from 6 to 40 feet, as may be necessary to give proper speed to the machinery they are designed to run—the buckets and nozzles being adapted to meet the conditions of head and power requirement.

1024

Duplex Air Compressor—Direct-driven by Pelton Wheel

Note the extremely simple construction and adaptation of the PELTON principle to direct-connection

A Novel Hydraulic Compressor Plant

The great flexibility of the PELTON system is exemplified by the plant installed at the Morning Mine, Mullan, Idaho. The problem here presented was unique in many ways, requiring that all the available water power be developed to best advantage in order to obtain the requisite output.

The plant consists of a 1000 horse-power duplex air compressor with three PELTON WHEELS mounted direct on the 15-inch shaft connecting the two sides of the compressor. This shaft is about 35 feet long, and is carried in four heavy babbitted pedestal bearings.

The water available is limited, and in order to obtain the required power it was necessary to utilize three different streams, affording heads of 1430 feet, 1150 feet and 140 feet respectively.

The water from both the 1430 and 1150-foot heads is applied to a single wheel 33 feet in diameter, which diameter is a compromise between the two high heads—taking 80 revolutions of the wheel as the average speed. This 33-foot wheel is the largest tangential water wheel ever made, and in order to withstand the severe strains to which it is subjected it required special construction. The wheel, which is mounted in the center of the compressor shaft, is of structural steel, built up, riveted and bolted, being entirely without cast parts, except the hub, which weighs about 12,000 pounds. The spokes are all in tension, and each is susceptible of adjustment in order to true up the rim, to which are bolted the buckets, which are only five inches by seven inches in dimensions—being extremely small in comparison to the wheel diameter. The total weight of the wheel is about 32,000 pounds, and it serves as a fly-wheel for the compressor. The water impinges on the wheel through two separate nozzles, which are mounted one over the other, each being supplied by a separate pipe line.

On either side of the large wheel is situated a 12-foot PELTON WHEEL, each of which has three nozzles and operates under a head of 140 feet. Each jet of the nozzles is controlled by a stream cut-off, all of which are connected together by suitable levers and links so that they operate in succession, thus maintaining the highest possible degree of efficiency.

Automatic governing is accomplished with these wheels only, the high-pressure nozzles maintaining a constant load on the large wheel. The fluctuations in pressure due to changes in load are extremely small and gradual, and a mercury column was adopted as the most sensitive device for operating the governor. The mercury column is arranged to balance the air pressure of 100 pounds, and closes or opens an electric circuit, which, operating on magnetic clutches, actuates the cut-offs on the triple nozzles, thus varying the capacity of the plant to meet the demands for air at the mine.

Particular attention is called to the wide variations in the different heads employed and to the manner of adapting them all to one speed by the simple methods of changing the wheel diameter. All belting and gearing losses are thus avoided, and the maximum efficiency is obtained.

Waipori Falls Electric Power Company, Limited—Hydro-electric Unit

There were two units furnished the above company of New Zealand. This is a typical example of "double overhung" construction. The engine-type generator is of 1000 k.-w. capacity, at 429 r. p. m. Maximum capacity of each unit, 2000 horse-power. Head, 674 feet.

Pelton Wheels and Pumps

An interesting application of the PELTON WHEEL to pumping engines is that illustrated in the Niagara Falls pumping plant. The plant consists of two complete Riedler duplex pumps and PELTON WATER WHEELS.

The wheels and pumps are located in a subterranean chamber excavated in the solid rock 130 feet below the Niagara River, the tail water discharging through a tunnel below the Falls.

The pumps supply water to the city at 70 pounds pressure, and therefore operate at 124 pounds normally, the back-pressure being 54 pounds. They are also arranged to supply 140 pounds pressure in the city mains for fire service when required, at short notice.

The PELTON WHEELS operate under an effective head of 125 feet, and run at a normal speed of 76 revolutions per minute. They are 12 feet in diameter, and are made with a fly-wheel rim weighing 12,000 pounds. To this are bolted the requisite size and number of buckets. The wheels are mounted directly on the pump shaft, all belting and gearing thus being avoided. The water is supplied through a specially constructed double nozzle, one of the water jets being used for normal, and both for fire pressure.

Each wheel is supplied with an automatic controller to prevent the pumps running overspeed, and, although all of the machinery is located in the pump chamber, the whole is under the control of the station operator in the power house 130 feet above.

The water for the wheels is controlled by a hydraulic gate operated by a three-way valve at the ground surface, the system being regulated and pressure in the mains kept constant by a Bourdon gauge, which actuates electro-magnets to either turn the water on or off the wheels. The switchboard is provided with two of these gauges, one regulating for the normal, or 70 pounds, and the other for the fire, or 140 pounds pressure. By operating a switch one or the other of the Bourdon gauges is thrown into action, which causes the wheel to bring the pump up to and hold it at the required pressure. The whole arrangement is exceedingly simple and easy to operate, and may be considered an ideal arrangement where water power is available for operating city water works pumps.

The Riedler pumps are duplex double-acting, each with a capacity of 6,000,000 gallons in 24 hours, against 70 pounds pressure. The pumps take water from the filters, into which it has been discharged by electrically-driven pumps—there being nine filter tanks. These are cleansed by flushing with a 5-inch stream from the pump mains, which occasions a great demand on the 14-inch main; yet so close is the regulation that the reduction in pressure is extremely small and lasts but a few seconds.

Electric Power Transmission

It is conceded that the electric transmission of energy is now one of the most important factors in the industrial world. The development on these lines has been of phenomenal growth, and the achievements of today can scarcely be credited from the standpoint of a few years ago. The world's first successful experiment in long-distance electric transmission was made in Germany in 1891, where a small amount of power was transmitted from Lauffin to Frankfort, and used in connection with an industrial exhibit held there at that time. In less than a year thereafter a PELTON WHEEL was driving a generator at the plant of the San Antonio Light and Power Company of Pomona, California, and transmitting electricity for power and light a distance of forty-five miles. California and the PELTON WHEEL thus claim the first power transmission plant in the United States, and the second in the world. In fact, the German installation was of an experimental nature, and the Pomona plant may be said to have been the first commercially operative system of electric transmission that was ever installed.

The development was rapid from that time on, until today a transmission of one hundred miles is of common occurrence, and thousands of horse-power are being transmitted for a distance of nearly three hundred miles. It has been demonstrated that the limit of distance to which power can be economically transmitted depends solely on the cost of the line conductors; hence it is difficult to forecast the development which is still to come. It is, after all, the utilization of water power which has made possible this great revolution, as there is no other power agency that will afford sufficient economy of production to make transmission of power commercially practicable. With water as the motive force, under all ordinary conditions electricity is conceded to be the most economical and reliable power known.

PELTON WHEELS meet so fully the existing requirements of this service as regards high efficiency, close regulation and small cost of maintenance, that they have come to be regarded as factors of prime importance in modern water power installations. The system is so flexible that it admits of adaptation to all conditions and every variety of service, and in so simple a way as to provide against liability of accident or interruption to continuous service.

The adaptation of electric power transmission must always be determined by the conditions of each particular case. Every proposition of this character is an engineering problem in itself to be carefully considered and worked out after a full investigation of all the facts and circumstances connected with it.

On the following pages will be found a list embracing 431 electric power installations operated by PELTON WHEELS, aggregating some 454,890 horse-power, which will in itself evidence the fact that this Company has specialized in the electrical field, and has had a wide experience. As will be observed, these plants are running under a great variety of conditions as to head, speed and power, yet in every instance, so far as known, they have proven to be efficient and reliable, affording an economic and satisfactory power.

Electric Power Installations

For Whom Installed	Horse-power	Operating Head Feet
Portezuelo Electric Power Company, Mexico	4000	461
Ned. Ind. Mijnbouw My., Celebes, East Indies	650	550
Petropolis Electric Light Company, Brazil, S. A.	1000	260
Cie de Boa Vista, Diamantina, Brazil, S. A.	400	350
Cripple Creek District Railway, Cripple Creek, Colo.	400	550
Burmah Ruby Mines, Mandalay, India (third station).	200	60
Miller Manual Labor School, Albemarle, Va.	70	225
Diamond Hill Gold Mining Company, Townsend, Mont.	700	170
Duplantier Electric Light Company, Costa Rica, C. A.	150	52
San Ildefonso Paper Mill, San Ildefonso, Mexico	1000	185
Yuba Electric Power Company, Marysville, Cal.	2000	290
Crested Butte Light and Water Company, Colorado	75	300
Development Syndicate, Butte County, Cal.	183	120
Kyoto City—Electrical Department—Kyoto, Japan	150	100
Alumbrado El de Quezaltenango, Guatemala, C. A.	140	83
South Yuba Water Company, Newcastle, Cal.	134	440
Concheno Mining Company, Concheno, Mexico	260	190
Santa Ysabel Mining Company, Jamestown, Cal.	80	130
Nevada County Electric Power Company, Cal.	1600	200
Juneau Electric Light Company, Juneau, Alaska	200	225
Hilo Electric Light and Power Company, Hawaii, H. I.	260	250
Tuolumne Electric Light and Power Company, Cal.	500	995
Oroville Gas and Electric Company, Oroville, Cal.	75	100
Cooperative Mining and Milling Company, Arizona	75	150
Empress Electricio de la Antigua, Guatemala, C. A.	200	65
Spring Creek Electric Power Company, Shasta, Cal.	300	800
Telluride Power Transmission Company, Telluride, Colo.	900	901
Cañon Creek Electric Power Company, Gem, Idaho	180	90
Columbia & Western Railway Company, British Columbia	550	267
Sandon Water Works and Light Company, British Columbia	200	400
Fort Wayne Electric Power Company, Arizona	260	200
Waianae Electric Company, Hawaiian Islands	270	690
Boca Ice Company, Prosser, Placer County, Cal.	140	20
Payson Electric Light and Power Company, Utah	150	125
Gold Hill Water Company, Virginia City, Nevada	130	230
Big Dipper Mining Company, Iowa Hill, Cal.	120	230
Gold Bluff Mining Company, Downieville, Cal.	125	270
Pioneer Mining Company, Plymouth, Cal.	125	560
Gold Dredging Company, Bannock, Mont.	150	350
Caroline Mining Company, Colorado (second station)	520	650
Ouray Electric Light and Power Company, Colorado	350	250
Hidden Treasure Gold Mining Company, Cal.	200	810
Jumper Mining Company, Stent, Cal.	400	230
Mountain Copper Mining Company, California	400	240
Ontario Silver Mining Company, Park City, Utah	160	120
Antigua Electric Light Company, Guatemala, C. A.	280	65
Silver Lake Mining and Milling Company, Colorado	300	180
Santa Fe Water and Investment Company, New Mexico	120	160
Santa Gertrudis Mining Company, Orizaba, Mexico	250	100
Arawaka Mining and Milling Company, Japan	250	100
Empresa Electrica Antigua, Guatemala, C. A.	260	65
Los Compania Electrica, Medillin, U. S. Colombia	600	490
Cia Electrica San Cristobal, Venezuela, S. A.	100	150
Cia de Luz Elec. de Heredia, Costa Rica, C. A.	400	200
San José Electric Light Company, Costa Rica, C. A.	400	200
Buttermilk Falls Electric Company, New York	200	85
Ophir Mining Company, Ophir Hill, Utah	100	108
Alumbrado Electric Company, C. A.	750	100
Neihart Water Company, Neihart, Mont.	175	310
Tjimpaka Tea Estate, Island of Java, D. E. I.	170	60
Dutch East Indian Electric Light Company	800	560
Bear Valley Electric Light Company, Nova Scotia	110	90
Santa Ana Electric Company, San Salvador, C. A.	200	60
Talemanco Electric Light Company, Venezuela, S. A.	160	200
Mendoza Electric Light Company, U. S. Colombia	110	76
Bridgetown Electric Light Company, N. S. W., Australia	140	126
Rossland Electric Light Company, Rossland, B. C.	250	240
British Columbia Electric Railway, Victoria, B. C.	1200	570
Blue Lakes Water Company, Blue Lakes, Cal.	2000	1050
Odawara Electric Railway Company, Odawara, Japan	1000	450
Redlands Electric Power Company (second station)	800	600
Ventura Land and Power Company, California	125	65
Salt Lake & Ogden Railway Company, Utah	175	230
Griffith Con. Mining Company, Eldorado County, Cal.	500	1100
Central California Electric Company, Auburn, Cal.	750	200
Bell Electric Company, Auburn, Cal.	100	150

Electric Power Installations—*Continued*

For Whom Installed	Horse-power	Operating Head Feet
Pike's Peak Power Company, Colorado	3000	1180
Golconda Mining Company, Limited, Oregon	200	380
Buffalo Mill Company, Buffalo, Wyoming	100	454
South Bend Electric Company, Washington	150	400
Redlands Electric Power Company, Cal. (fourth station)	400	550
Gold Note Mining Company, Arizona	300	160
Clipper Mining Company, Pony, Montana	150	110
Rawhide Mining Company, California	1800	1200
Northport Electric Company, Washington	125	175
Ferrocarria Electrica de Jalapa, Mexico	1200	244
La Hormiga Transmission, Contreras, Mexico	800	860
Cia Electrica e Yrrigadora Tetapango, Mexico	3600	195
Oregon Lumber Company, Oregon City, Oregon	165	135
Reno Water, Land and Light Company, Nevada	520	34
San Poil Mining and Water Power Company, Wash.	185	105
Nelson Electric Company, British Columbia	200	185
San Miguel Power Transmission Company, Ames, Colo.	1000	580
Colorado Electric Light and Power Company, Colorado	180	223
Redlands Electric Light and Power Company, Cal. (third station)	350	605
Kanagawa Electric Light Company, Japan	191	200
Four Hills Mining Company, Plumas County, Cal.	300	910
City of Healdsburg Electric Light Company, Cal.	200	963
E. Basadre Forero, La Paz, Bolivia, S. A.	150	230
Lihue Plantation Company, Kauai, Hawaiian Islands	125	45
Elling & Morris Mining Company, Pony, Montana	150	108
Fern Gold Mining Company, British Columbia	190	250
Granite Bimetallic Con. Mining Company, Montana	1800	680
Morning Mining and Milling Company, Mullan, Idaho	240	900
Utica Mining Company, Angels, Cal.	1100	536
Weaver Mining Company, Ballarat, Cal.	100	200
Butte County Electric Power and Light Company, Cal.	1200	600
Sunnyside Mining Company, Eureka, Colo.	200	261
Yellowstone Mining Company, British Columbia	150	430
Moendjoel Estate, Tjibadak, Java, East Indies	275	100
Cia Minera y Beneficiadora de Teziutlan, Mexico	1220	1050
Canas Mining Company, Guernavaca, Mexico	140	70
Fabrica La Experiencia, Guadalajara, Mexico	1400	460
Cia de Boa Vista, Diamantina, Brazil, S. A.	125	170
Cia Luz Electrica de Orizaba, Mexico	1200	320
Fabrica Santa Teresa, Contreras, Mexico	300	270
The Basic Mining and Milling Company, Idaho	800	350
Noriega, Sanchez y Cia, Los Molinos, Mexico	720	110
Burmah Ruby Mines, Limited, Burmah, India	500	200
Hale Electric Light Plant, Dixville Notch, N. H.	150	360
Cia Gas Acetylino, Peru, S. A.	400	125
Telluride Power Transmission Company, Colorado (second station)	1600	550
Central California Electric Company, Cal. (second station)	1200	430
Truckee River General Electric Company, Cal.	2300	84
Sierra Power Company, Southern California	1100	640
Crystal Lake Gold Mining Company, California	450	900
Stanislaus Power Company, Calaveras County, Cal.	1400	140
Montauk Consolidated Gold Mining Company, Cal.	100	115
Seattle Light and Power Company, Washington	200	230
Glenwood Light and Power Company, Colorado	100	460
United Electric, Gas and Power Company, California	100	350
Colorado Electric Light and Power Company, Colorado	100	223
Kanagawa Electric Light Company, Japan	600	200
Cape Colony Electric Power Company, South Africa	1300	560
Trinidad Electric Light and Power Company, W. I.	1050	460
Cia Industrial de Santa Catalina, Mexico	1400	70
Eagles Mere Electric Light Company, Pennsylvania	250	220
Cornucopia Mining and Milling Company, Oregon	400	320
Annie Laurie Mining and Milling Company, Utah	450	550
Mauritius Electric Company, Island of Mauritius	100	150
Frontino & Bolivia Gold Mining Company, S. A.	150	120
La Hormiga Transmission Company, Mexico (second station)	400	860
Teziutlan Copper Mining Company, Mexico	900	1050
Santa Ysabel Mining Company, Tuolumne County, Cal.	260	30
Keswick Electric Power Company, Shasta County, Cal.	4500	1200
Samoa Estates, Limited, Samoa Islands	160	30
Kofu Electric Power Company, Japan	200	90
Bagnall & Hilles, Gomei Kaisha, Japan	250	100
Wanoosnock Electric Power Company, Massachusetts	600	130
British Columbia Electric Railway Company, B. C. (second station)	900	650
Big Creek Power Company, Cal. (second station)	1500	975
Regla Electric Power Transmission Company, Mexico	3000	800

Electric Power Installations—*Continued*

For Whom Installed	Horse-power	Operating Head Feet
Big Cottonwood Power Company, Salt Lake City, Utah	3000	380
Folsom Electric Power Company, Folsom, Cal.	4000	55
Nevada County Electric Power Company, California	1000	210
Santa Ysabel Mine, Tuolumne County, Cal	500	300
Tuolumne County Electric Company, California	500	950
Gold Valley Mining and Milling Company, California	250	200
Boza Electric Power Company, Venezuela, S. A.	1200	400
Big Creek Electric Power Company, Santa Cruz, Cal.	800	840
Redlands Electric Power Company, Redlands, Cal.	1000	500
Petropolis Electric Power Company, Brazil, S. A.	800	260
Quezaltenango Electric Company, Guatemala, C. A.	250	55
Ontario Mining Company, Park City, Utah	300	120
Alaska Treadwell Mine, Douglas Island, Alaska	600	460
Colorado Springs Contract Company, Colorado	440	600
Silver Lake Mining and Milling Company, Colorado	700	180
Roaring Fork Electric Power Company, Aspen, Colo.	1250	330
People's Electric Light and Power Company, Colo.	700	180
Telluride Electric Power Company, Colorado	1000	500
Caroline Mining and Milling Company, Colorado	400	500
Mount Morgan Mining Company, South Africa	600	120
Hilo Electric Light Company, Hilo, H. I.	250	260
Walla Walla Electric Company, Washington	750	60
Amecameca Electric Light and Power Company, Mexico	700	980
Nelson Electric Light and Power Company, B. C.	350	160
Juneau Electric Light and Power Company, Alaska	200	108
Bucaramanga Electric Light Company, Colombia, S. A.	400	53
Kyoto Electric Power Company, Kyoto, Japan	1000	110
Chollar Mining and Milling Company, Nevada	750	1680
San Antonio Electric Power Company, California	800	400
Standard Consolidated Mining Company, Bodie, Cal.	650	340
Couer d'Alene Silver Mining Company, Idaho	760	810
Belmont Consolidated Mining Company, Colorado	250	610
Mammoth Mine, Madera County, Cal.	175	60
Glenwood Light and Power Company, Colorado	450	380
Casapalca Electric Light Company., Casapalca, Peru	400	170
Electric Light and Power Company, Costa Rica	400	200
Mount Lowe Railway Company, Altadena, Cal.	200	1250
Revenue Tunnel and Mining Company, Colorado	600	650
South Yuba Canal Company, Newcastle, Cal.	130	420
Central California Electric Compnny, Newcastle, Cal.	1200	420
Roaring Fork Electric Light and Power Company, Colo.	1400	820
Bell Electric Company, Auburn, Cal., (second station)	225	140
Cia de Luz Electrica, San Salvador, C. A.	300	60
Santa Ana Electric Company, San Salvador, C. A.	400	76
Medillin Electric Light Company, U. S. Colombia	700	340
Cia Esplotadora de Loto y Coronel, Chili, S. A.	650	360
F. D. Mendiola, Boza, Costa Rica, C. A.	400	200
Cartago Electric Light Company, Costa Rica, C. A.	300	250
Moodies Mining Company, Limited, South Africa	800	130
Honolulu Electric Light Company, Honolulu, H. I.	100	200
Bozeman Electric Light Company, Montana	170	124
Wallace Electric Light Company, Wallace, Idaho	125	124
Bell Electric Light Company, Auburn, Cal.	100	80
Alaska Gold Mining Company, Alaska	150	460
Banner Mining Company, Butte County, Cal.	160	120
Cooperative Mining and Milling Company, Arizona	100	150
Helena and Livingston Smelting Company, Montana	600	725
Burmah Ruby Mines, Mandalay, India (second station)	400	120
Fairhaven Electric Company, Fairhaven, Wash.	120	300
Phoenix Mining Company, New Zealand	200	180
Bear Valley Electric Power Company, Nova Scotia	270	170
Tinebi Electric Light and Power Company, C. A.	300	260
Weaverville Electric Light and Power Company, Cal.	700	200
Mullan Electric Light and Power Company, Idaho	160	170
Calumet Mining Company, Shasta County, Cal.	300	800
Delmatia Mining Company, Eldorado, Cal.	230	110
Gold King Mining and Milling Company, Colorado	1200	500
Sheridan-Belmont Mining Company, Colorado	360	240
Barrio-Nueva Jute Company, Orizaba, Mexico	700	100
Southern California Power Company, Cal.	5000	700
Cia de Papal de San Rafael y Anexas, Mexico	1200	950
Cia de Papal de San Rafael y Anexas, Mexico	550	220
Hiroshima Electric Company, Hiroshima, Japan	1200	240
Utah Power Company, Salt Lake City, Utah	2000	440
San Gabriel Electric Company, California	3200	400
Tuolumne County Water Company, Cal.	1500	925

Electric Power Installations — *Continued*

For Whom Installed	Horse-power	Operating Head Feet
Big Creek Power Company, California	900	923
Chainman Mining and Electric Company, Nevada	180	177
Montana Electric Light and Power Company, Montana	150	680
Utica Gold Mining Company, California	1000	530
Lucky Girl Mining Company, Nevada	100	660
Winnemucca Water and Light Company, Nevada	150	1100
Mariposa Commercial and Mining Company, Cal.	730	29
Oroville Electric Light and Power Company, Cal.	300	243
Roseburg Water Company, Oregon	150	240
Wenatchee Electric Light and Power Company, Wash.	150	160
Los Gatos Ice and Power Company, California	150	216
Cia de Transmission Elec. de Potencia, S. A.	1800	800
Ardjasarie Electric Power Transmission Company, Java	100	282
Goldpan Engineering and Mine Supply Company, Colorado	540	460
Yosemite Valley Lighting Plant, California	250	144
Eagle-Shawmut Mining Co., California	100	900
Bay Counties Power Company, California	2100	292
Standard Electric Company, California	1800	1000
Hilo Electric Light Company. Hawaiian Islands	240	360
Bishop Light and Power Company, California	100	50
Clark Electric Light and Power Company, Utah	400	700
Ouray Electric Light and Power Company, Colorado	700	400
United Light and Power Company, Colorado	560	720
Lahaina Plantation, Hawaiian Territory	400	565
Bay Counties Power Company, California (second station)	1400	590
Fraser & Chalmers, London, England	200	400
Toccoa Falls Light and Power Company, Georgia	150	245
Pike's Peak Power Company, Colorado (second station)	1150	600
Mexican General Electric Company, Mexico	100	48
San Simonito Power Developing Company, Mexico	700	1650
Angel Sanchez & Brothers, Mexico	250	350
Standard Electric Company, California (second station)	3000	1500
Utica Electric Light and Power Company, New York	125	255
Jalapa Light and Power Company, Mexico	900	250
Bay Counties Power Company, California (third station)	1400	590
Orizaba City Lighting Plant, Mexico	320	600
Yellowstone National Park, Wyoming	600	230
Caucasus Copper Company, Limited, Russia	500	180
Edison Electric Company, California	1000	1900
Ophir Hill Con. Mining Company, Utah	600	836
Hazel Gold Mining Company, California	300	800
Lewiston Electric Power Company, Idaho	500	215
Vancouver Power Company, British Columbia	10000	400
Pike's Peak Hydro-electric Company, Colorado	5000	2150
Nephi City Electric Power Plant, Utah	150	100
Northern California Power Company (second station)	6600	1150
Rock Creek Power and Transmission Company, Oregon	1760	942
American River Electric Company, California	6600	572
Archbishop Gillow, Sinaloa, Mexico	480	115
Societe Industrial de Sta. Catalina, Peru	1500	155
Guanajuato Power and Electric Company, Mexico	6600	320
Allis-Chalmers Company, South Africa	320	360
Mexican Light and Power Company, California	100	310
Cloverdale Light and Power Company, California	350	218
Big Springs Electric Company, Utah	250	270
Siskiyou Electric Power Company, California	1100	680
Brigham City Power Plant, Utah	800	280
Republic Light and Power Company	225	172
Aomori Electric Light Company, Japan	450	320
Fukushima Electric Company, Japan	600	260
Springville Power Plant, Utah	200	140
Gaston Gold Mining Company, California	260	500
Silver Cup Mining Company, British Columbia	520	130
Columbus Consolidated Mining Company, Utah	660	494
Utah County Light and Power Company, Utah	500	285
Seattle Municipal Plant, Seattle, Wash.	5600	550
Cia Aviadora de la Mina Natividad, Mexico	500	254
Empresa Electrica de Santa Rosa, Peru (second station)	1500	155
Marconi Transmission Plant, Mexico	425	592
Edison Electric Company, California (seventh station)	1700	305
Wenatchee Electric Company, Washington	360	760
Utah County Light and Power Company (second station)	500	285
Puget Sound Power Company, Washington	31000	865
Washington & Oregon Power Company, Wash.	3000	356
North Mountain Power Company, California	1100	600
Beaver City Municipality, Utah	200	125

Electric Power Installations—*Continued*

For Whom Installed	Horse-power	Operating Head Feet
Waipori Falls Electric Power Company, New Zealand	4000	655
Britannia Copper Syndicate, Limited, British Columbia	1000	1900
Silver Lake Mines, Colorado	250	365
Treasury Tunnel Mines, Colorado	150	180
Honolulu Municipality, Hawaiian Territory	500	390
Nanaimo Electric Light and Power Company, B. C.	450	159
Los Gatos Ice and Power Company, Cal. (second station)	150	168
Mitsui & Company, Japan	300	350
Casapalca Mining Company, Peru	500	540
Teziutlan Copper Company, Mexico	450	1050
Abangarez Gold Fields, Limited, C. A.	200	260
A. Saenz Barranquilla, South America	150	410
Carbonear Electric Light Company, N. F.	325	210
Animas Canal Reservoir Power Company, Colorado	8000	970
Cia de Luz Electrica, Mexico (second station)	1750	460
Topaz Mining Company, Nicaragua	260	150
J. W. Beckwith, Nova Scotia	175	210
Petropolis Electric Light Company, Brazil (second station)	200	192
Guaratingueta Electric Light Company, Brazil	300	130
Sao Paulo Electric Company, Brazil	1000	75
Bagnall & Hilles, Japan	200	180
British Columbia Electric Railway Co. (second station)	2000	621
Nevada Power, Mining and Milling Company, Cal.	3000	1064
La Grande Water Storage Company, Washington	800	580
Grande Ronde Electric Company, Oregon	600	887
Makee Sugar Company, Honolulu	500	390
Bandoeng Electricitat, Mattschappy, Java	570	258
Monarch Construction Gold Mining Company, Colo.	150	520
Kauai Electric Company, Hawaii	5000	575
Mount Whitney Power Company, California	2700	1290
Siskiyou Electric Power Company, California	2000	715
D. G. Aguirre, Mexico	1200	175
Cumberland Electric Light Company, B. C.	100	280
Nevada Power, Mining and Milling Company, Cal.	3000	990
Homestake Mining Company, South Dakota	800	470
Shibaura Engineering Works, Japan	500	580
Butler & Company, Alaska	150	450
Ephraim City Municipal Plant, Utah	400	486
Pagilaran Estate, Java	150	212
Mines of Huayna Potosi, South America	200	430
Gaston Gold Mining Company, California	100	569
Takata & Company, Japan	350	200
Oro Water, Light and Power Company, California	3400	465
Lewiston-Clarkson Company, Washington	500	220
Vancouver Power Company, B. C. (fourth station)	3000	360
United Light and Power Company, Colorado	900	692
Columbus Consolidated Mining Company, Utah	660	494
California Gas and Electric Corporation, California	10000	765
Northwest Light and Water Company, Washington	1500	142
Yukon Consolidated Goldfields, British Columbia	2400	650
United Light and Power Company, Colo. (second station)	550	650
Yukaichi Electric Light Company, Japan	700	630
Alaska Electric Light and Power Company, Alaska	600	225
Fremont Power Company, Oregon	2200	1092
Shibaura Engine Works, Japan (second station)	600	580
Wellington Colliery Company, British Columbia	100	467
Trinity Bonanza King Mining Company, California	1100	250
Telluride Power Company, Colorado (fourth station)	4000	1800
Telluride Power Company, Colorado (fifth station)	5000	900
Cumberland Electric Light Company, B. C.	250	975
Haines Electric Power Company, Oregon	110	75
Utah County Light and Power Company, Utah	2800	540
Spring City Electric Light Company, Utah	100	280
E. J. Baldwin, California	300	126
Fairbanks, Morse & Company, Washington	200	240
New York Grass Valley Gold Mining Company, Cal.	100	98
Cia de Luz Electrica, South America	500	66
Ventanas Consolidated Mining Company, Mexico	800	400
Makaweli Sugar Company, Hawaii	100	276
Oroville Light and Power Company, California	275	240
Amori Electric Company, Japan	450	320
Amsinck & Company, South America	400	176
Takata & Company, Japan	380	370
Stanislaus Electric Power Company, California	37200	1490
Northern California Power Company, Cal. (third station)	16750	360
Vancouver Power Company, B. C. (fifth station)	10500	360

Electric Power Installations—*Continued*

For Whom Installed	Horse-power	Operating Head Feet
Mitsui & Co., Japan	500	345
Dolores Mines Company, Mexico	250	844
Skagit Improvement Company, Washington	250	590
Parowan City Corporation, Utah	100	110
Cedar City Light and Power Company, Utah	150	163
Superior Portland Cement Company, Washington	2400	422
Pacific Power Company, California	250	110
Wallace Light and Water Company, Idaho	800	310
Cia de Luz de Comitan, South America	200	205
Quincy Electric Light and Power Company, California	175	360
Oro Water, Light and Power Company, California	1500	240
City of Los Angeles, California	1500	600
Bungo Traction Company, Japan	600	305
Steptoe Valley Smelting and Mining Company, Nevada	200	465
Kekaha Sugar Company, Territory of Hawaii	900	275
Summit County Electric Power Company, Colorado	1000	450
Nevada-California Power Company, California	7340	928
Northern Light and Power Company, California	3000	721
Societe Anonyme, South America	150	430
Northern California Power Company, Cal. (fourth station)	4000	1100
Steptoe Valley Smelting Co., Nevada (second station)	400	421
Eccles & Browning, Utah	700	1010
Ruby Gulch Gold Mining Company, Montana	165	185
Mexican Light and Power Company, Mexico	130	300
Dalton Power Company, Massachusetts	850	152
Quito Electric Light Company, Ecuador	325	190
New York & Honduras Rosario Mining Company	300	834
Tramway Light and Power Company, Brazil	400	230
D. F. Payne, New York	700	290
Fraser & Chalmers, Limited, England	725	537
Prefectura de Pocos de Caldos, Brazil	310	275
Empresa Electrica de Santa Rosa, S. A.	1260	150
Cia de Papal, Mexico	650	1250
Tanaka Gold Mines, Japan	260	340
La Hormiga Mills, Mexico	400	492
Lake Dunmore Power and Transmission Company, Vermont	210	115
Donnadieu, Veyan & Company, Mexico	750	700
Empresa Electrica de Santa Rosa, Peru	5000	240
Portezuelo Electric Light and Power Company, Mexico	1750	460
Quito Electric Light Company, Ecuador (second station)	325	190
United Towns Electric Company, Northfield (second unit)	325	210
Blue Mountain Electric Company, Pennsylvania	106	350
Peruvian Mining, Smelting and Refining Company, Peru	180	120
Anglo-Mexican Electric Company, Limited, Mexico	120	400
Teziutlan Copper Company, Mexico	450	1050
Takata & Company, Japan	380	370
Gabriel Mancera, Mexico	265	475
Mining Exploration Company, South America	275	485

Summary of Pelton Wheels Now in Use

Some idea of the extent to which PELTON WHEELS have come into use may be obtained from the following list. There are now running considerably more than 12,000 PELTON WHEELS in various parts of the world in connection with mining, manufacturing and other industries, aggregating in excess of 1,300,000 horse-power.

In the United States and Foreign Countries	Number of Wheels	Aggregating Horse-power
California, Oregon and Nevada	8554	836,053
Washington, Idaho and Alaska	762	76,607
Utah, Colorado and Montana	241	36,166
Hawaii, New Mexico and Arizona	244	18,114
Middle, West and Atlantic States	183	9,842
Mexico and Central America	707	172,279
Various South American States	421	27,360
Australia, New Zealand, Japan and India	549	50,591
East and West Indies Islands	268	34,857
British Columbia and Nova Scotia	123	39,350
England and South Africa	102	13,855
Germany, France, Italy and Spain	406	6,096
Norway, Sweden and Denmark	44	2,996
Total	12604	1,324,167

NOTE.—California is credited with by far the largest number because the PELTON WHEEL was invented and first introduced in that State, and for the further reason that water is abundant there for power purposes and under favorable conditions as to head.

1026

Northern California Power Company — Interior View

Three 1500 horse-power PELTON WHEELS, direct-connected to 750 k.-w. electric generators.   Head, 1150 feet

NOTE.—The above Company has now installed and under construction a total of 15 PELTON WATER WHEEL Units, aggregating 32,300 horse-power

The Pike's Peak Power Company — Interior View of Station
(For description, see opposite page)

1027

The Pike's Peak Power Company

On the opposite page is shown a partial interior view of the transmission plant of this company, located near Victor, Colorado. This installation is particularly interesting from a hydraulic standpoint—embracing what is said to be the first steel-faced, granite back-filled dam in the United States. This structure has a length of 375 feet across the cap, with a cross-section of 16 feet, and 210 feet along the base, the extreme thickness of which is 138 feet.

The pipe line has a total length of approximately 22,000 feet, and consists of part wood and part sheet steel pipe. For a distance of 19,000 feet, to a point involving a static head of 170 feet, the water is carried by a wood stave pipe 30 inches in diameter, made of California redwood, banded with one-half inch steel bands, spaced at such intervals as are necessary for resisting the internal pressure. The wood stave pipe joins a sheet steel riveted pipe 29 inches in diameter, of thickness varying from three-sixteenths to three-quarters of an inch, as required to stand the pressure involved, with a safety factor of four. The total length of the steel pipe is 3000 feet, laid on an incline averaging 38 per cent.

The station consists of four 400 k.-w. electric generators with their exciters, all direct-connected to Pelton Water Wheels, operating under an effective head of 1160 feet. Each wheel unit consists of two cast-steel disc Pelton Wheels 66 inches in diameter, keyed on the same shaft and working in the same wheel housing. While the full power from each unit could have been readily obtained from a single wheel, two wheels are used for the purpose of obtaining greater flexibility in the plant and highest efficiency at partial load. As will be noted from the illustration, the base frames are of the same type and general design as the generators to which they are connected. The frames of the water wheels and generators are accurately faced and are rigidly connected to each other by bolts and dowels.

The connection of the water wheel and generator shaft is effected by a 6500-pound cast-iron fly-wheel, banded with a rolled steel tire four inches in thickness. This wheel is seven feet in diameter, and is keyed to the wheel shaft. On the generator shaft is keyed a hub carrying a face coupling which is bolted to the fly-wheel, thus forming an accurate and rigid connection of the shafts of the two machines.

The nozzles are of the ball and socket deflecting type, operated by rock-shafts and levers. The nozzle-tips used on the eight wheels are of varied diameters, so proportioned that by making combinations on the various units, the required power at almost any stage of load can be obtained with the deflecting nozzles in their normal position; thus the maximum efficiency is obtained at all times. The gate valves are of special design, being of the Pelton single disc type, and provided with roller bearings, which render them easily operated by hand under full pressure without the intervention of gearing.

The advantages of hydraulic power in connection with electric transmission will be appreciated when it is considered that this plant is very successful commercially, in the face of the enormous expenditure necessary in the construction of the huge dam and laying of the long pipe line under the severe conditions presented by the topography of the country.

Hydro-electric Power Plant

Two PELTON WHEEL units, each of 2500 horse-power maximum capacity, direct-connected to 1500 k.-w., 300 r. p. m., engine-type generators. Also two 100 horse-power exciter units. Head, 575 feet

1028

The Stanislaus Electric Power Company

The hydro-electric installation of the above company is of interest on account of the magnitude of its operations and the extreme conditions involved.

The power house is situated on the Stanislaus River, about fourteen miles from the town of Angels, Calaveras County, California. A diverting dam was built on the Stanislaus River, and the water carried by a flume and ditch line 7900 feet long, of which about 1500 feet of the lower end is ditch. The flume is 6 feet 6 inches high and 9 feet wide, inside measurement, and is designed to carry 500 second feet. The flume construction differs slightly from the common practise in that it is built on 15-foot bents instead of the standard 8-foot bents. It is very heavily timbered to carry the additional weight falling on the stringers. Further improvement over standard practise lies in the interlocking and bolting together of the box posts with the sills which support the box.

The water is thus carried to a forebay reservoir located on a hill above the power house. This forebay has a depth of approximately 50 feet, and an available capacity between high and low water levels of approximately 13,000,000 cubic feet. When drained to the lower level, there will still be sufficient head over the entrance to the pipe lines to provide for velocity and entrance with approximately 4 feet additional to spare. The forebay is constructed on two sides of a saddle, thus actually forming two reservoirs, the upper one of which serves as a settling basin for the water before it passes from the saddle into the lower basin, from which the pressure pipes are supplied.

From the forebay to the power house the water is carried by a pipe system consisting of two lines, cross-connected 1300 feet from the forebay and again at the power house, where they merge into a header from which separate distributing pipes are led to each unit. Each of these pipe lines, beginning at the reservoir, consists of a 66-inch riveted flanged steel pipe line, 300 feet long, under the reservoir; 1300 feet of 66-inch wood stave pipe; 1300 feet of 48-inch and 1700 feet of 38-inch inside diameter riveted steel pipe, to the power house header. The sheet steel pipe is of various thicknesses, from $\frac{1}{4}$-inch at the junction of the wood stave pipe to 1 inch thick at the extreme lower end.

The material for both the wood staves and the flume was all cut and milled on the ground, the wood staves being made of carefully selected sugar pine, and the flume of both sugar and yellow pine, the former predominating. The wood stave pipe is subjected to a maximum pressure due to 100-foot head. The static head from forebay to power house is 1495 feet, which, at normal load, will give an effective working head on the wheels of 1400 feet.

The initial equipment consists of three PELTON WATER WHEELS, each direct-connected to a 6700 k.-w., 400 r. p. m., alternating current, three-phase, 60-cycle, engine-type generator; also two PELTON WHEELS for driving 600 horse-power exciter units direct-connected. Each main PELTON unit has a maximum overload capacity of 12,000 horse-power. Thus, the total present capacity of the station is 37,200 horse-power with provision in all branches for an ultimate maximum development of 80,000 horse-power. The water wheel units are of the PELTON "double overhung" type, the chief advantages of which are the handling of large units of power without undue proportion of bucket and nozzle dimensions, and a minimum of floor space required. This construction, briefly, consists of one PELTON WHEEL overhanging each end of the shaft beyond the bearings, of which there are two for each unit: these are located one on either side of the engine-type generator, which is in the center, its rotor being mounted on the water wheel shaft. This construction enables the bearings to be more equally loaded than if the entire output was obtained from one wheel. Also the fractional load efficiency is higher and the regulation closer with this design.

Each wheel unit and each exciter unit is controlled by a PELTON self-contained oil pressure governor actuating the needle-deflecting nozzle — all subject to pilot control at the switchboard.

The initial voltage of the main units is 4000, from which it is stepped up on the line to 114,000 volts, current being delivered at the sub-stations at approximately 100,000 volts.

The ultimate length of the transmission line will be approximately 140 miles, with San Francisco as the destination. The line is supported by suspension insulators on steel towers with an average span of about 850 feet. Each insulator consists of five separate elements, and each element was tested electrically to 90,000 volts, and mechanically to 5000 pounds strain, before it left the factory. The five combined elements have a tested strength in excess of 300,000 volts.

The contract with this Company for the hydraulic equipment commenced at the header pipe, and included the furnishing of the heavy cast-steel "Y's" and all connecting pipes to the wheels.

As of interest to engineers are given below *pro forma* specifications covering one complete unit, representing, it is claimed, the latest design and refinement of impulse wheel construction :

Specifications

General

Head	1475 feet effective.
Horse-power . .	37,200 maximum overload.
Units	Three main and two exciter units.
Type	Double overhung — one pair of wheels to each main unit. Rotor mounted in center of shaft. Exciter units single overhung direct-connected.
Speed	Main units, 400 r. p. m. Exciter units, 720 r. p. m.

Specific

Wheels	Two for each main unit.
Diameter	Suitable for a normal speed of 400 r. p. m.
Material	Forged and cast-steel and nickel oil tempered steel.
Details	Steel discs to be of open-hearth steel containing not to exceed .4 per cent carbon and .08 per cent phosphorus. Test bars to show a tensile strength of not less than 60,000 pounds per square inch. Discs to be turned perfectly true on all surfaces and balanced up to a runaway speed of 600 r. p. m. Buckets to be of cast-steel, straddling and bolted to the disc by means of fluid-compressed nickel steel bolts with hexagon semi-finished nuts, the bolts being ground and hydraulically

Specifications — *Continued*

pressed into place. Bucket castings to be thoroughly annealed, with inner surfaces machined and fine ground to true curves.

Lugs of buckets to be milled with standard cutters to a forced fit on disc rim. Each bucket to be accurately centered so that the cutting edges shall be exactly in the same central plane. Buckets all to be balanced and brought to uniform weight so that each wheel will be in dynamic and static balance. All drilling, reaming and other machine work to be to standard templet to secure absolute interchangeability of parts.

Main Shaft . . . To carry rotor of generator and both PELTON WHEELS overhanging the bearings.

Dimensions . . . 25 feet over all; 20 inches at center where rotor is carried; 16½ inches at journals and 14 inches at the overhang.

Material . . . . To be made from a single ingot of fluid-compressed 3½ per cent nickel steel, hollow forged throughout and oil tempered.

Details . . . . A test piece cut from shaft forgings must show a tensile strength of not less than 105,000 pounds per square inch; an elastic limit of 79,000 pounds; an elongation of not less than 24 per cent, and a contraction of area of not less than 54.67 per cent. Shaft to be turned to rough gage, then keyseated and returned to lathe for final finish and polish, thus eliminating any spring due to keyseating.

Main Journal Bearings . . . . Two for each unit.

Dimensions . . . 16½ inches diameter by 60 inches long.

Type . . . . . PELTON ring oiling generator type with ball and socket.

Details . . . The main body of bearing to be of heavy cast-iron, carefully machined and bored out for a cast-iron shell; the shell to be of heavy construction, lined with highest grade babbitt peined in, then bored out and finally scraped to a uniform and exact bearing surface for the journal at running temperature and grooved for the passage of oil; final thickness of babbitt to be not less than ⅜ of an inch at any point. Four oil rings provided for throwing oil. Bottom half of shell to be removable when relieved of shaft weight without shifting the generator; the shell to be held in place by a cap bolted to bearing. Each end of bearing provided with sight gages to show the level of oil in reservoir. Lower half of bearing to be water-jacketed throughout by means of grilled water compartments next to the babbitted surfaces. Bearing also to be provided with oil-cooling ducts connecting to an oil pumping system by which forced lubrication under pressure

Specifications — *Continued*

may be made; also provision for oil emergency flushing. Bearing shells to be provided with oil catching grooves, and drip outlets cut in each end to catch oil flowing out along the shaft. Bearing housing to project over shaft shoulders so as to catch oil thrown therefrom. Bearings to be provided with finished glass oil gages and finished brass turn-cocks.

Gate Valves . . .   Two for each unit.

   Material . . . .   Cast-steel annealed.

   Diameter . . .   20 inches.

   Type . . . . .   PELTON single steel disc with removable bronze seat, outside screw and yoke, rising spindle.

   Details . . . .   Valve stem of nickel steel; nut of gun metal; provided with oil packed, steel roller bearings to take thrust in each direction. Valve provided with gear and worm wheel for quick and slow motion by hand, and equipped with reversible PELTON MOTOR for quick operation. 4-inch by-pass provided with extra heavy connection. All joints flanged and pressure connections made with forged steel bolts; nuts and heads faced. Round rubber gaskets to be used, and all joints machined to gage to facilitate matching. Valve arranged for packing under pressure.

   Test Requirements   Maintain a cold water pressure of 1300 pounds per square inch for a period of five hours.

Nozzles . . . . .   Two in number (one for each wheel).

   Diameter . . . .   20 inches; flanged to match PELTON gates.

   Type . . . . .   Combined needle and deflecting with ball and socket joints.

   Details . . . .   Main castings to be perfectly sound and free from blow holes. Trunnion pins of highest grade nickel steel forgings working in renewable gun metal, oil packed bearings. Trunnion bolts provided with forged outside reinforcing straps. All bolts on nozzle of special steel forgings turned and faced under head and nut. Swinging portion of nozzle, weighing 6700 pounds, to be counterbalanced by a hydraulic cylinder taking pressure from main pipe. Nozzle to have proper connections for operating by governor. Needle to be of nickel steel with means for the adjustment of the position of needle and size of stream by hand; also by a direct-current reversible motor operated by a double-throw switch while unit is in operation; needle to be securely held and positively locked in any position of adjustment; needle portion to be turned perfectly true to curves of least resistance, and to be free from flaws and highly polished. Control of the needle to be gradual and easy of operation; and an indicator provided showing the nozzle opening for all positions.

Specifications — *Continued*

Test Requirements

Maintain a cold water pressure of 1300 pounds per square inch for a period of five hours.

Wheel Housing .

Each wheel to be enclosed in a ½-inch steel plate upper housing, supported on cast-iron foundation base frame; upper portion to be easily removable and provided with eye-bolts for handling. Housings to be caulked and made perfectly water-tight; also internally braced to prevent vibration. Housings to be provided with PELTON patented centrifugal discs and pockets where shaft passes through, to prevent leakage. Covered openings to be provided for inspection of wheels. Housings to be equipped with spatter boxes with which to catch spent water; this water to be conveyed by piping to and through the bearing shell and oil cellars of the bearings; a part of the spatter water to be delivered into the generator end of the hollow shaft. All exposed piping and fittings for carrying water to be of polished brass with finished brass valves and fittings.

Bed-plate . . . .

To be of cast-iron and incorporated with the bearing sole-plates, and the extension of the generator sole plate.

Exciter Units . .

To be supplied two sets of exciter wheels operating under generally similar hydraulic conditions to the main wheels; each set to consist of one PELTON WHEEL mounted on exciter generator overhung shaft; gate valve and cross-connecting pipes; one housing and one deflecting nozzle with hand regulating needle and governor.

Accessories . . .

One PELTON oil pressure governor with necessary operating connections for each unit, stationary tachometers, pressure gages, brass railings for each unit, floor plates and pipe connections to make a finished installation.

Material Test . .

Where not otherwise provided, materials are to be tested in accordance with the rules of the International Association for testing materials, all test pieces being taken from coupons attached to each particular part; test pieces for buckets to be taken from each heat.

Guarantees . . .

The material and workmanship used in this construction are guaranteed to be of the highest grade, free from flaws or defects of any kind; any parts proving defective within two years from date of starting to be replaced free of charge by the manufacturer.

Sufficiency . . .

It is the intention of the above specifications to describe a complete water wheel unit ready for operation. Any parts which may not have been specifically mentioned and which are necessary for the operation and complete assembly of the plant, to be supplied.

The Vancouver Power Company

The above company, operating also the British Columbia Electric Railway Company controls one of the largest hydro-electric systems in the Northwest. The principal power station is located on Burrard Inlet, an arm of the Puget Sound, about 16 miles from Vancouver, British Columbia. The water supply is obtained by damming a cañon. which drains an immense watershed area; in addition a tunnel 12,760 feet in length connects with a large lake located on a different watershed, which acts as a balancing reservoir, the flow being controlled by sluice gates.

The water is conveyed from the dam, a distance of 1800 feet, to the power house, there being a separate pipe line for each unit, consisting of part wood stave and part sheet steel riveted pipe, affording an effective head on the wheels of 390 feet.

As will be noted from the illustration above, the power house is located on the bay shore, so that the water, after passing through the wheels, discharges directly into tide-water.

The installation at present consists of four 1500 k.-w. and one 5000 k.-w. engine-type generators, all running at 200 r. p. m., and each driven by PELTON WATER WHEELS, direct-connected; in addition are two exciter units, also direct-connected. The combined capacity of this station at present is 23,300 horse-power, and provision is made for a substantial increase at a later date. The PELTON WHEEL units are of the double overhung, iron-mounted type, provided with automatic governors.

The electric power is transmitted 17 miles to the sub-station at Vancouver, and at one point crosses an arm of the inlet by a cable span 2800 feet in length, supported by a steel tower on either side.

The British Columbia Electric Railway Company's station is located a few miles from Victoria, British Columbia, where are installed four PELTON WHEEL units, operating under a head of 625 feet, the capacity of the station being 4000 horse-power. Thus the entire system controls at the present time in excess of 27,000 horse-power. A ready market is found for this power at Vancouver, Victoria, and the surrounding towns, particularly in connection with the suburban electric roads.

Puget Sound Power Company — Interior View

Four wheel units, aggregating 30,000 horse-power, in this station, of the double overhung type, coupled to 3500 k.-w., 225 r. p. m. generators Each unit has an overload capacity of 7500 horse-power. (For description, see page 86)

1029

The Puget Sound Power Company

Located on the Puyallup River, 32 miles from Tacoma, Washington, the installation of this company is one of the most important in the Northwest.

The Puyallup River has its origin in the glaciers and snow peaks of Mount Rainier, the highest mountain in the United States. As a consequence, an unfailing source of water is assured from the melting snow and ice.

The water scheme consists of diverting the Puyallup River and carrying its flow by means of a flume 10 miles long to a reservoir located on a high plateau, and thence by steel pipes to the PELTON WHEELS, affording a head of 865 feet. The flume and reservoir are constructed with a view to the ultimate development of 60,000 horse-power, and the present equipment consists of four direct-connected PELTON WHEELS, each driving a 3500 k.-w. generator at 225 r. p. m. Also two PELTON WHEELS, each direct-connected to 150 k.-w. exciters. Each wheel unit has an overload capacity of 7500 horse-power, making the output of the station 30,000 horse-power.

As will be noted from the interior view of this station on page 85, the wheel units are of the PELTON double overhung type with the rotor of the generator in the center. Combination needle and deflecting nozzles are used, the governing mechanism and gate valves being controlled from the switchboard. On the opposite page is illustrated an exterior view of this station, showing the discharge water from the wheels in operation.

This power is transmitted to Tacoma and to Seattle, 48 miles distant, being used for the various industrial enterprises in that section, and particularly for operating the extensive system of suburban electric roads in the vicinity of Seattle.

The contract with this company for the hydraulic equipment commenced at the terminals of the four main pipe lines, at which point are installed heavy cast-steel " Y " pipes with extra large ribs and special flanges.

1030

Stream of water from PELTON needle nozzle operating under 390-foot head and developing 1500 horse-power. Note the shadow of needle showing through stream, and the perfect form of jet, which is as shown in original photograph.

Puget Sound Power Company — Exterior View

Showing wheels in operation

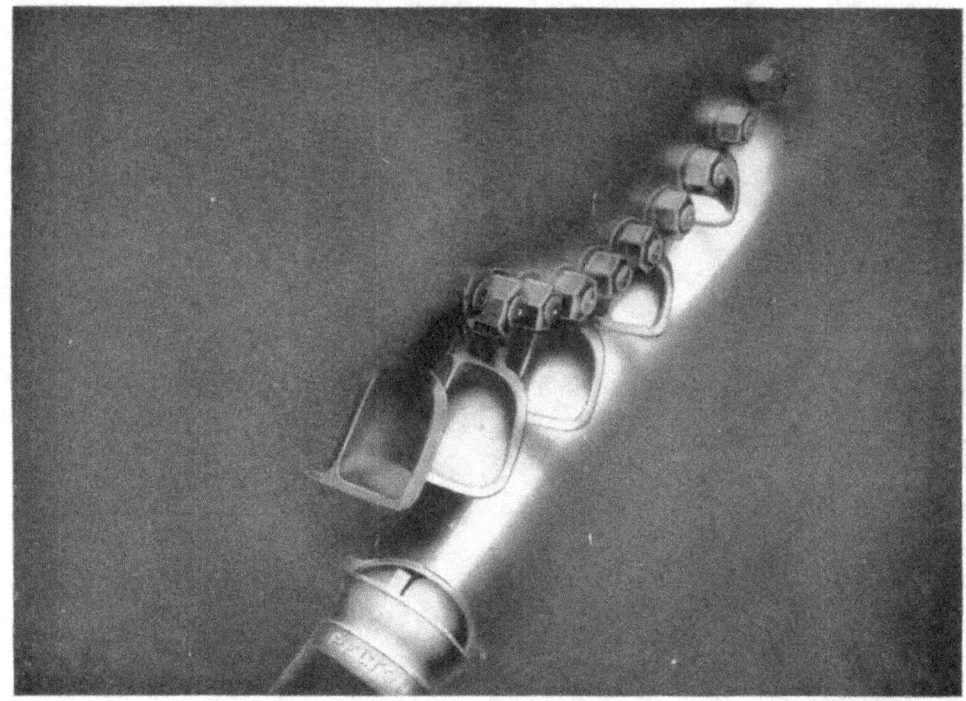

1032

Instantaneous Photograph of Stream of Water from a Pelton Needle Nozzle
Operating under an extremely high head

A Notable High Head Installation

The Pike's Peak Hydro-electric Company of Colorado Springs, has the distinction of operating a water wheel plant under the highest head available in the United States. In fact, there is but one other installation in the world utilizing a higher head, and that for a small amount of power.

The plant in question is located on the outskirts of the town of Manitou, Colorado, and consists of three PELTON units, each direct-connected to a 750 k.-w. electric generator running at 450 r. p. m. The net head on the PELTON WHEELS is 2150 feet, equivalent to the enormous pressure of 935 pounds per square inch.

The wheels are mounted in the pulley compartment of the generator, and are provided with combination needle and deflecting nozzles operated by hydraulic governors. The gates, nozzles and other pressure parts are of cast-steel, designed with a large safety factor, and were subjected to a cold water test of 2000 pounds per square inch before installing. The wheels proper consist of cast-steel discs with gun metal buckets, fine ground and machined inside. Each wheel has an overload capacity of 1500 horse-power.

Some idea as to the severe strain to which the apparatus is subjected may be gained by noting that the water issuing from the nozzle-tip under the head of 2150 feet has a spouting velocity of 22,300 feet per minute—in excess of 250 miles per hour. This means that the wheel buckets, which travel at practically one-half of the spouting velocity of the water, must stand a centrifugal strain equal to a speed of 125 miles per hour.

Current is transmitted to Colorado Springs for power and lighting purposes and is also largely consumed by the mines and mills in that vicinity.

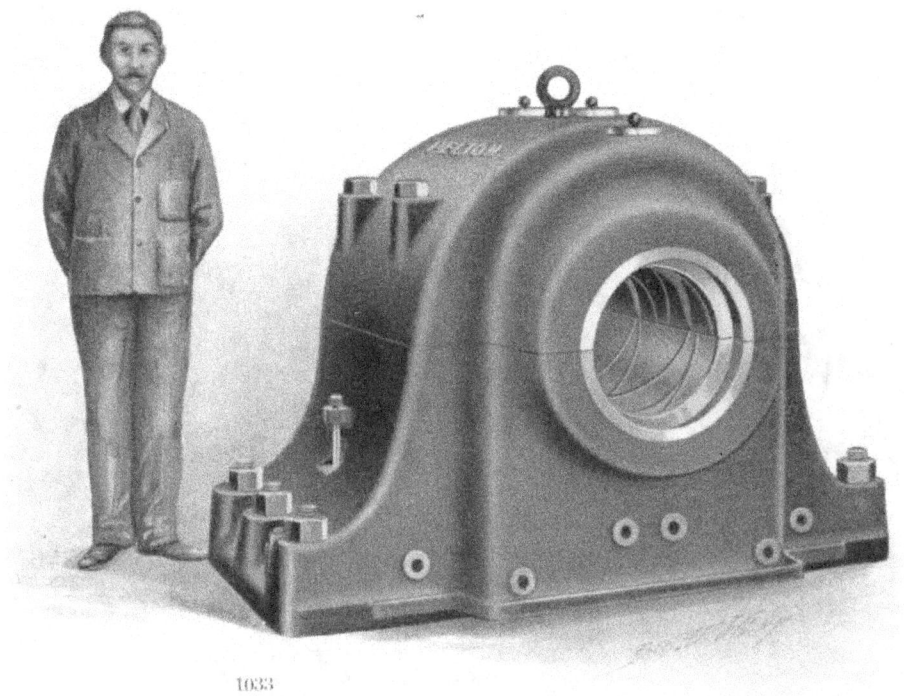

1033

16½-inch Journal Bearing

(For specifications, see page 81)

Mechanical Construction of Pelton Wheels

Water wheel apparatus as the prime mover must be *reliable*—that is to say, it must be capable of affording absolutely continuous service. Reliability is dependent largely on three factors: proper adaptation of material, correct design and proportion, and high grade of mechanical workmanship.

The material entering into the construction of PELTON apparatus is carefully tested for inherent defects; wheels and accessory parts are designed with a large safety factor to provide against undue strain. PELTON apparatus is the result of the best mechanical workmanship it is possible to obtain. As an illustration may be given a brief description of some of the details entering into the mechanical construction of PELTON WHEELS.

BEARINGS: This Company manufactures various kinds of bearings and pillow blocks —ring-oiling, ball and socket, and rigid—for use in connection with water wheel line-shafting, engine-type generators, direct-connected compressors, etc. Particular attention is called to the PELTON ring-oiling bearing shown on page 90, and to the special bearing illustrated above. This design of bearing corresponds throughout with that on the highest grade of electrical apparatus; barrels are lined with the best of special babbitt, peined in, bored, scraped to a true fit at running temperature, and oil grooves properly cut. The journal is flooded with oil by means of heavy bronze rings rolling on the shaft, and running in oil contained in a large oil reservoir. Suitable drain cocks and sight glasses are fitted in the pedestals. The babbitted barrel is turned to a true

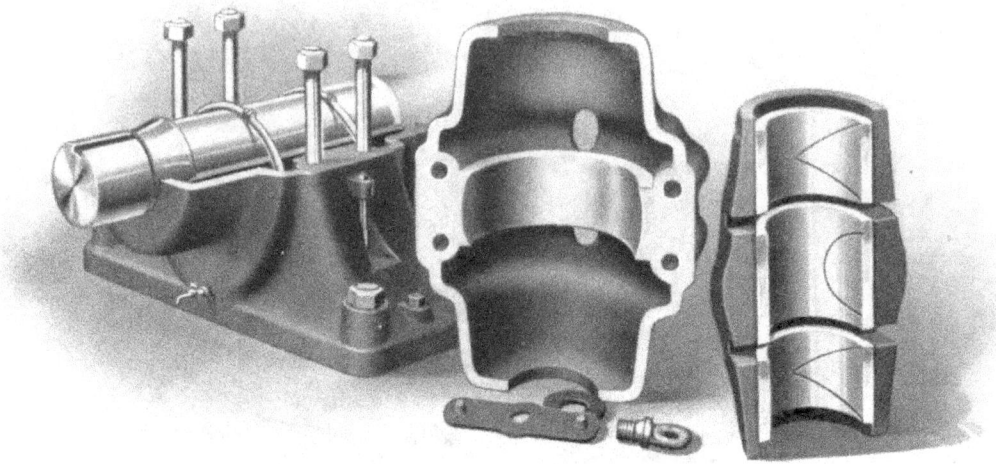

1034

1035

Pelton Automatic Ring-oiling, Ball and Socket Journal
Showing details and assembled bearing

spherical surface fitting in the concave surface of the main pedestal stand. These surfaces are not babbitted together, as on the cheaper constructions, but form a close fitting joint, turned and scraped to an accurate fit. By relieving the barrel of the weight of the shaft, the lower half may be easily removed. The bronze rings are made parting so that they may also be removed while the shaft is in place.

These journals are also incorporated in pedestal stands, forming GENERATOR-TYPE BEARINGS, harmonizing in shape with those on generator. They are made of any height, and to take any diameter of shaft. (See description on page 81 of special journals for high duty.)

WHEEL CENTERS are of steel or cast-iron, depending on existing conditions, and for heavier powers are of the disc type, turned all over, for balance.

The BUCKETS for this type are milled to an exact radius and straddle the periphery of the center, being secured to same by means of turned steel bolts in reamed holes. Accurate jigs and templets are preserved, to admit of exact duplication in the event of possible accident, to which all machinery is liable. Buckets are constructed of steel, phosphor-bronze or cast-iron, as may be best suited to the requirements. Wheels are forced on the shaft by hydraulic pressure and securely keyed, making slippage impossible. All wheels are accurately adjusted to a running balance.

All SHAFTING is first turned to approximate gage, then keyseated and afterward returned to the lathe for final cut and polish, thus eliminating any possible spring due to keyseating, and insuring true running; shafting is provided with oil and thrust collars.

1036

Flexible Leather Link Coupling

For direct-connection of generator and water wheel shafts

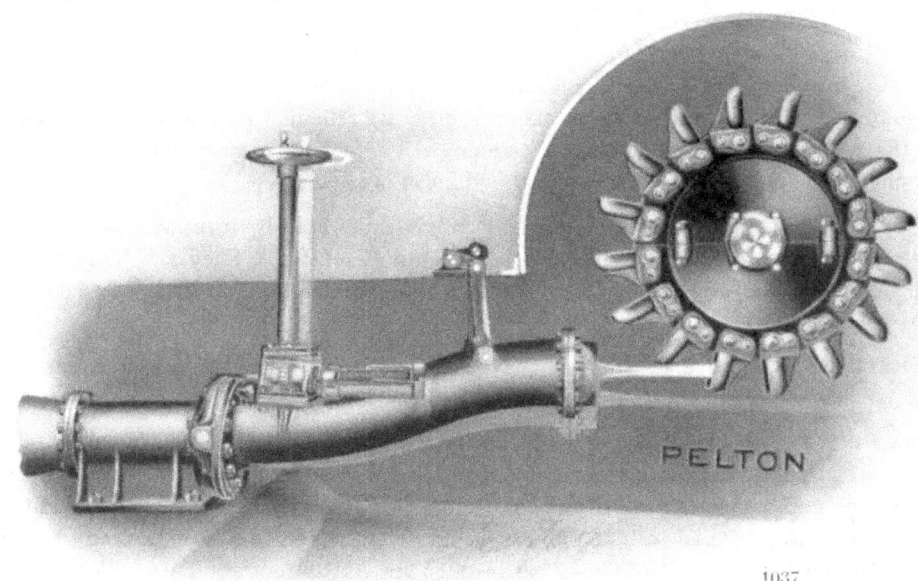

1037

Pelton Needle and Deflecting Nozzle for Economic Regulation
(See description of operation on page 13)

NOZZLES are of steel or cast-iron, depending on the head involved. They are divided into three classes: rigid, plain deflecting, and the combination needle and deflecting nozzle. The rigid nozzle, as its name implies, is secured and stationary, and is used where no regulation is required, or where stream cut-offs are employed (see page 12). The plain deflecting nozzle is used where regulation is essential without water economy. The combination needle and deflecting nozzle is for close regulation and extreme water economy (see page 13).

In addition to the above a further refinement has been devised in the PELTON automatic needle nozzle. This involves the principles of the combination needle deflecting nozzle with added mechanism, by which the position of the needle is *automatically* and gradually changed to correspond with the fluctuation in the power load and the positions of the deflecting nozzle.

With this device the services of the operator for adjusting the needle position may be entirely dispensed with, this function being performed through the medium of a small governor, the result being that the water quantity used is in almost direct proportion to the power output at all times, regardless of the fluctuation of the power load. Additional information regarding the automatic needle nozzle will be furnished on request.

BED-PLATES, where used, are of most substantial construction, heavily ribbed to secure rigidity and accurately planed on all matching surfaces.

HOUSINGS enclosing wheel compartment are of cast-iron or sheet steel—caulked and made water-tight, and with cast-iron planed flanges for joints. All are provided with PELTON patented centrifugal discs and pockets to prevent leakage of water along the shaft.

GATE VALVES: Special descriptions and illustrations of the gate valves manufactured by this Company will be found on pages 95 to 99.

The various illustrations in this catalog, all of which are from photographs of machinery actually constructed, indicate the design and finish of this Company's product, which, it is claimed, are in accordance with the best engineering practise.

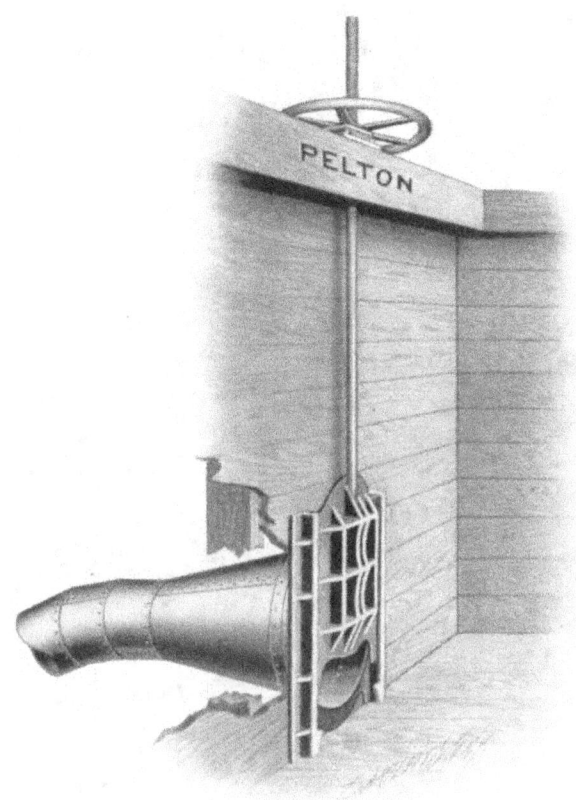

1038

Entrance Taper and Head Gate in Flume

Water Wheel Accessories

This Company makes a specialty of manufacturing all the various accessories necessary in a water wheel and pipe installation. Above is shown a HEAD GATE for use in a flume. This is advisable in a long pipe line so that the pipe may be emptied to admit of repair to the wheels or pipe without draining the ditch or flume. With the head gate are often furnished "grizzly" and screen for preventing leaves and trash from entering the pipe. The head gate can also be adapted for masonry or earth dams.

AIR VALVES should be placed at the high places in a pipe line, so that they may open and admit air when the pipe is drained—otherwise the atmospheric pressure is liable to cause the pipe to collapse.

STRAINERS are generally recommended when hydraulic governors are used in power plants, and are absolutely essential if the water carries sand or grit which cannot be readily settled. The mechanism of the hydraulic governor is necessarily so delicate that gritty water will soon cut the valves and seats, and render it inoperative, for which reason the oil pressure governor is usually recommended.

PRESSURE GAGES to indicate the static and running pressure are advisable in any pipe line. Gages reading in pounds or feet head, or both, are manufactured expressly for this Company, and guaranteed accurate.

1039

Battery of Safety Relief Valves

Above is illustrated a battery of SAFETY RELIEF VALVES. These are necessary, particularly in pipe systems operating under extreme pressures, and are usually placed at the lower end of the pipe line near the nozzle. The valves are set to operate at a pressure slightly greater than the normal, and in the event of the water flow being suddenly checked by the closing of the gate or operation of the governor, the safety valves momentarily open and relieve the pressure, thus guarding the pipe against the possibility of water hammer. These valves may be placed singly, or in a battery, as shown, depending largely on the size of pipe involved and the working head.

The PELTON mechanically-operated relief valve is an added protection to a pipe system, and is considered more reliable in extreme cases than any other form of device for accomplishing the same purpose. This is essentially a by-pass valve operating in connection with the governor, and is automatically opened as the governor responds to a decrease in power, and a consequent throttling of water on the wheel. The mechanism is such that the by-pass opens quickly and closes slowly, thus affording positive relief from excess pressure, at the same time eliminating the danger of water hammer, and *securing* the maximum of *water economy*.

This Company also manufactures a double strainer, consisting of two chambers (one in reserve), containing a series of screens. When one set becomes clogged the water may be diverted to the other cylinder, and at the same time a portion of the water will pass through the clogged screens in a *reverse* direction, thus removing all obstructions.

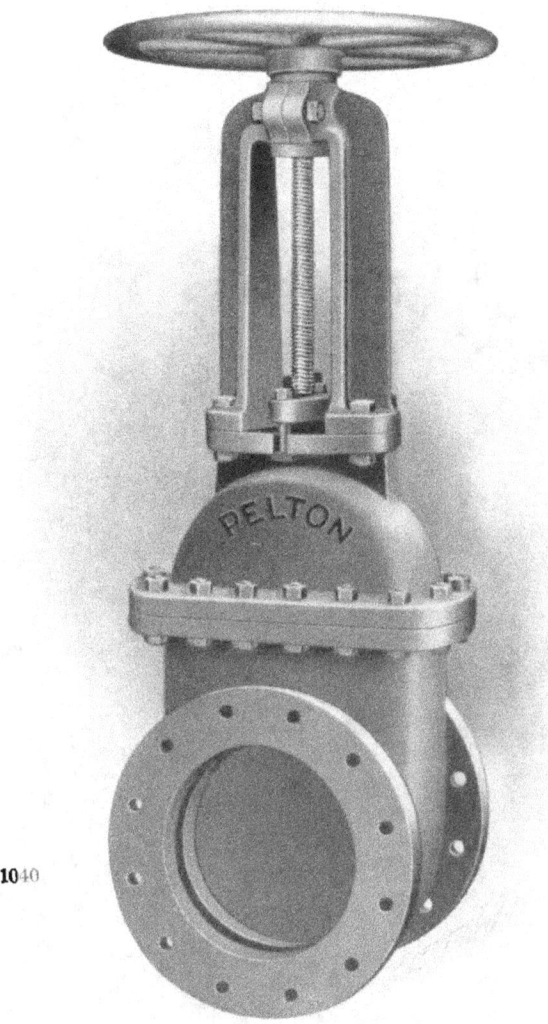

1040

Standard Type
Pelton Gate Valves

This Company makes a particular specialty of the design and manufacture of high-grade gate valves for all conditions of service, its product being the result of many years' experience. In general, PELTON gate valves are of the single disc, straight-way type with outside screw and yoke and rising spindle, illustration of the standard type being shown above. The disc rises clear of the opening, thus giving a perfectly free way for the water. The seats are of bronze and can be replaced when worn. The valve is provided with an outside screw spindle with threads of fine pitch, and a heavy yoke. Thus all operating parts are outside and readily accessible. They are furnished with flange, screw or hub ends for lead joints, and for use under any required pressure. For extreme pressures gearing or roller bearings are usually employed. These valves are made heavy and light for different ranges of pressure, and are manufactured in sizes from 4 inch upward. All valves are tested to twice the pressure for which they are intended. On pages 96 to 99 will be found illustrations of various types of valves manufactured by this Company to meet special conditions. Prices and specifications on valves for any service furnished on application.

Group of 20-inch All-steel Pelton Geared Gate Valves

Equipped with roller bearings; operated by reversible PELTON water motors. Tested to 1300 pounds per square inch
(For detailed specifications, see page 82)

57

Pelton 24-inch Steel Gate Valve

Single disc type—with outside screw and yoke and rising spindle. Arranged for operating by electric motor, and provided with roller bearings to take thrust from stem. Tested to 1000 pounds pressure.

24-inch Cast Steel Bronze Lined Pelton Gate Valve
Hydraulic actuated and provided with by-pass.   For 900 feet head

1041

Special High Pressure Steel Pelton Gate Valve

With by-pass and roller bearings on stem.   Tested to 2000 pounds per square inch

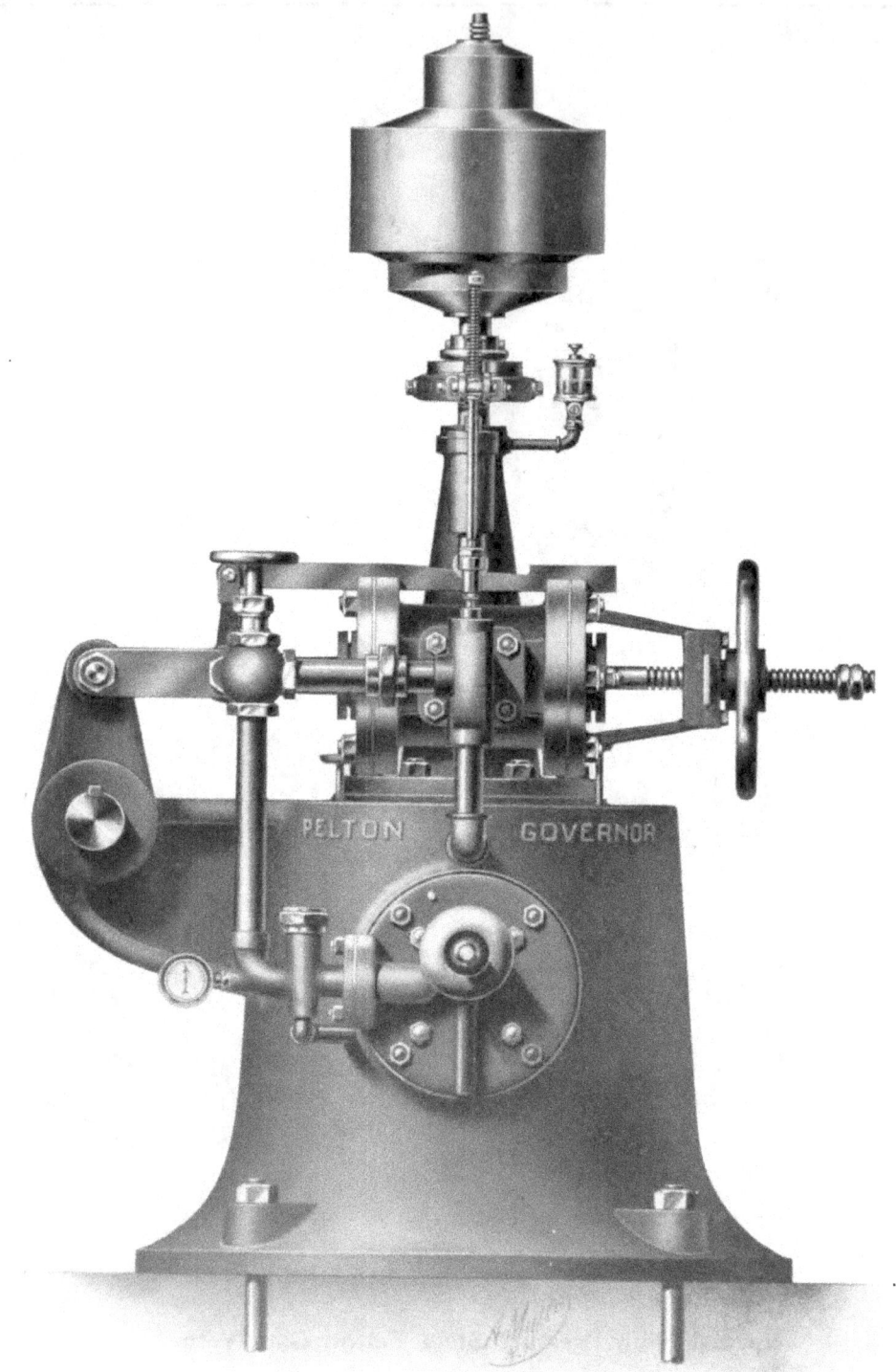

1042

Pelton Oil Pressure Governor
Self-contained type.   (For description, see page 101)

Regulation of Pelton Wheels

As explained on page 12 of this catalog, the speed control of the PELTON WHEEL is accomplished by the use of various devices. These are usually intended to be actuated by an independent governor, the operation of which is entirely automatic.

This Company manufactures an oil pressure governor for use in connection with PELTON WHEELS, a description of which follows :

The governor is made, broadly speaking, in two types · the SELF-CONTAINED, which has its own pump, forming part of the apparatus; and the SPECIAL TYPE, which is supplied with oil pressure from an independent pumping system.

The SELF-CONTAINED GOVERNOR, as shown in the cut on opposite page, has a heavy cast-iron base containing a gear pump and oil reservoir, with operating cylinder and controlling mechanism. The latter consists of a pair of heavy weights, set upright on knife edges and working against a spring resistance. The weights are driven by a vertical shaft to which they are connected by levers, and operate a collar sliding upon the shaft, this collar being connected to the regulating valve controlling the admission of oil to the cylinder. The governor is provided with a relay mechanism to prevent "hunting" or racing, and with suitable adjustments to permit of running under or over speed and synchronizing.

The special type is of much greater capacity than the self-contained, the general principles, however, being the same. This type is intended for use in large hydro-electric generating stations, where there are a number of units, all of the governors being supplied with the same pumping system.

In a hydro-electric station, regular and efficient service is, to a large extent, dependent on the governor. As a consequence, much care has been exercised in the design and mechanical workmanship of the PELTON governor. Ports, valves and other pressure parts are carefully ground to an exact fit, and the operating mechanisms are designed with an ample safety factor to provide against breakage.

Specific statements as to the degree of regulation possible, can only be made on receipt of complete data as to operating conditions, but it may be stated that this Company is prepared to install these governors with rigid guarantees for high performance under the most exacting conditions of load variation.

Parties desiring information as to the governing apparatus should state the character of machinery to be governed, its speed and the amount of fluctuation involved. If for electrical machinery, it should be stated whether it is for lighting or power load, or both. In all cases the head available and length and diameter of pipe should be given. If the inquiry relates to governor for use on an existing wheel, information should be given as to the character of controlling mechanism; that is to say, whether the wheel is equipped with a plain deflecting nozzle, combination needle and deflecting nozzle, or stream cut-offs. PELTON governors will be sold only for use in connection with PELTON WHEELS and PELTON-FRANCIS turbines.

101

1043

Self-contained Unit with Governor

The Pelton-Francis Turbine

This Company realized fully ten years ago that the great increase in the capacity of electric generators and, coincidently, the introduction of the steam turbine, with its extremely high speed and the consequent development of high-speed electrical apparatus, would call for an adaptation in water wheel construction to meet these changing conditions.   With this idea in mind, an exhaustive investigation was made with a view to the development of a water turbine which would fulfil the following requirements :

1.   A wide range in the head under which it could be operated, *viz.*, from 10 to 600 feet.
2.   A large capacity under practically any head within the limits above specified.
3.   A wide range of speed under all heads, as compared with capacity.
4.   Adaptability to close regulation under all conditions, irrespective of head and power.
5.   High efficiency under varying loads.

These conditions were best met by the Francis type of turbine, long favorably known in Europe, and this type, therefore, was decided upon : in order to distinguish it from the ordinary type of turbine, the name PELTON-FRANCIS turbine was adopted.

These turbines are built to run upon either a vertical or horizontal shaft, as may best suit the conditions in any particular case.   Where a vertical shaft is used, a single turbine is generally recommended, provided the required power and speed can be obtained, as a double turbine with a vertical shaft greatly complicates the design. In a horizontal shaft installation the turbine may be either single or double, as may best meet the requirements in any particular case.

The Runner

The PELTON-FRANCIS turbine is what may be best described as an inward flow reaction turbine; that is, the water flows from the rim, or outside, towards the center in practically radial lines, leaving the runner at an angle of 90 degrees from the direction of its flow. The runners are generally made of gun metal, and are cast in one piece; under low heads, cast-iron may be used.

The runners may be either single or double, and the shape varies greatly according to the conditions under which they are to be used. Each runner, is *specially designed* to meet the conditions under which it is to operate.

The Case

The turbine is generally enclosed in a spiral case, of either cast-iron or cast-steel, but under low heads, a steel plate case is used or the turbine may be set in an open flume. The spiral case permits of a much higher water velocity than would be possible with an ordinary cylindrical case, and at the same time, guides the flow of water in the proper direction, and eliminates "eddy currents." It also permits of the use of much smaller gate valves than would be possible with a cylindrical case.

The Gates

The gates control the admission of the water from the case to the runner. These are fixed about the periphery of the runner on pivots or shafts which project through ring covers or flanges bolted to both sides of the case, the draft tube or tubes being in turn bolted to the ring covers. The gates are made of either forged or cast steel, or gun metal. Their shafts in the small turbines have a bearing on one side only; on larger turbines the shafts of the gates project on each side, thus securing proper support. These shafts work in bronze bushings, with stuffing-boxes to prevent leakage. A lever is fitted to the end of each shaft and connected by means of bronze links and pins to a guide ring, by means of which all the gates are operated simultaneously (see illustration on page 107).

The above briefly describes the three most important parts, *viz.*, the runner, case and gates, as the proper design of these parts determines the efficiency and character of regulation of the turbine. A good gate mechanism, free from lost motion and easy to operate, is equally as necessary as is a good governor, where close regulation is desired.

PELTON-FRANCIS turbines are built on order only, each turbine being designed to meet a given set of conditions. Estimates furnished on these turbines in units of any size desired up to 25,000 horse-power.

The efficiency of the turbine depends largely upon the conditions under which it is designed to operate. Under favorable conditions efficiencies of 78 per cent at half load, 85 per cent at three-quarters load, and 82 per cent at full load may be obtained. Specific guarantees will be made in each case where there is a full understanding of the operating conditions.

Black Hills Traction Company's Station—Interior View

Two double PELTON-FRANCIS turbines, each 1050 horse-power capacity under 110 feet head. (For description, see page 105)

1044

The Black Hills Traction Company

The above company has installed at Spearfish, South Dakota, a hydro-electric station completely equipped with PELTON-FRANCIS turbines, some details of which will be of interest.

The water power development embraces a canal some ten miles in length, with a capacity of about 200 second feet of water, terminating in a forebay to which is connected a 78-inch wood stave pipe, about 4600 feet in length. The lower end of this pipe connects with a steel " Y " branch, 78 x 54 x 54 inches. Each 54-inch outlet is provided with a gate valve, and from each gate valve is led a 54-inch wood stave pipe about 840 feet long, each pipe supplying one main turbine unit.

A cross-connection, inside the power house, is made between the two pipe lines for supplying the exciter turbines; this is so arranged that the exciters may be operated from either or both pipe lines. About 100 feet above the " Y " branch, above referred to, there is a riser or standpipe, 44 inches diameter by 40 feet high.

The plant consists of two special 32-inch double PELTON-FRANCIS turbines, enclosed in spiral cast-iron cases. Each turbine has a maximum capacity of 1050 horse-power at 400 r. p. m., under 110 feet effective head, the static head from forebay being 117 feet. Each turbine is direct-connected to a 500 k.-w. generator, and is provided with a 16-inch relief valve, mechanically operated by the governor, arranged for quick opening and slow closing, thereby obviating the possibility of water hammer. In addition, there is a 10,000-pound steel banded fly-wheel 7 feet in diameter on the end of each turbine shaft, to which the coupling connecting the shaft of turbine and generator is bolted. The turbine runners and gates are of gun metal, and the regulating mechanism is outside the turbine case. Each main turbine is provided with a self-contained PELTON oil pressure governor actuating a rock-shaft, to which the gate ring is connected by double connecting rods.

The exciters are of 30 k.-w. capacity each, and run at a speed of 1000 r. p. m., each exciter being direct-connected to a 50 horse-power PELTON-FRANCIS, single, spiral case turbine, hand regulated.

The main turbines developed an efficiency of over 85 per cent under test, and the regulation is exceedingly close, the variation in speed with full load thrown off instantaneously not exceeding 3 per cent; the increase in pressure on pipe line was within 5 per cent.

Part of the power is transmitted to the Homestake Mine at Deadwood, South Dakota, and the balance is used in the vicinity of Spearfish.

The company has under consideration the construction of a second station some twenty miles distant from the present plant, in order to meet the growing demand for power.

The illustration on opposite page shows the general arrangement of station.

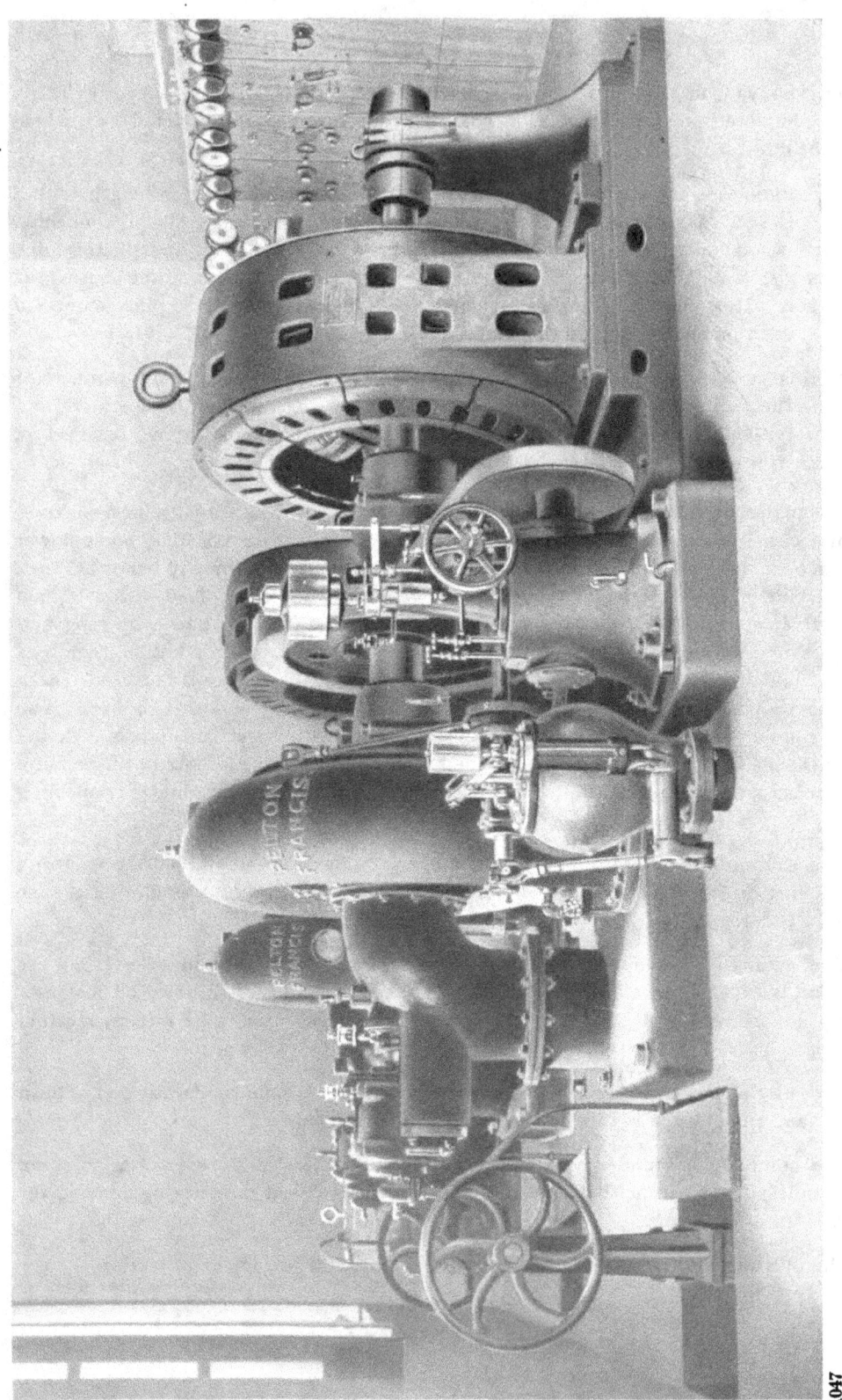

1047

Sultepec Electric Light and Power Company—Interior View

Two single discharge PELTON-FRANCIS turbines. 800 horse-power capacity each, at 900 r. p. m. (For description, see page 107)

1046

Dismantled View of Pelton-Francis Turbine—Showing Runner and Gates
(For description, see page 103)

The Sultepec Electric Light and Power Company

The plant of this company, located near Temascaltepec, Mexico, operates under an effective head of 340 feet, the water being supplied to the turbines through a steel pipe 42 inches in diameter by 950 feet in length.

The lower end of the pipe line terminates in a receiver 54 inches in diameter, to which are connected three special 24-inch PELTON-FRANCIS turbines, each having a maximum capacity of 800 horse-power at 900 r. p. m. Each turbine is direct-connected to a 450 k.-w. generator.

The turbines are single discharge, the draft tubes being about 20 feet in length. The runners and gates are of gun metal, and the water passages admitting water to the runners are bronze lined.

Each turbine is mounted on a cast-iron bed plate, and is provided with a steel banded fly-wheel, and a relief or by-pass valve mechanism operated from the gate ring by means of the governor. This is arranged for quick opening and slow closing, thus affording a safety device to protect the pipe from water hammer due to the sudden closing of the gates on account of short circuit or other extraordinary demand upon the governing mechanism. The turbines are controlled by self-contained oil pressure PELTON governors. Exciter current is provided by a 50 k.-w. direct-current generator driven by a PELTON WATER WHEEL.

Under actual test these turbines showed the remarkable efficiency of 84.4 per cent at three-quarters load, and 84.7 per cent at maximum full load. With the entire load thrown off instantaneously the variation in speed did not exceed 6 per cent, and the maximum increase in pressure at the lower end of pipe line was only 10 pounds, indicating the very efficient action of the by-pass valve mentioned above. Further information regarding the efficiency test is given on page 108. The cut on the opposite page shows the general arrangement of this station.

107

Efficiency Test of 24-inch Single Pelton-Francis Turbine

For The Sultepec Electric Light and Power Company

List	Time	Gate Opening Inches	Weir Inches	Water	Leakage	Water	Loss in Pipe	Effective Head	Volts	Amperes	Combined Efficiency Per Cent	Horse-power Input	Water Exciter	Water Main Turbine	Horse-power Theoretical	K.-W.	Generator Efficiency Per Cent	Horse-power on Wheel	Fly-wheel Horse-power	Effective Horse-power	Wheel Efficiency Per Cent
											Exciter										
1	11.55	0.683	11 5/8	14.64	0.23	14.41	0.25	339.	117	38	30	20.	0.52	13.89	532	274	90	408	10	418	78.
2	12.00	0.683	11 5/8	14.64	0.23	14.41	0.25	339.	117	38	30	20.	0.52	13.89	532	283	90	422	10	432	81.2
3	12.15	0.91	13 3/8	18.08	0.23	18.45	0.39	338.7	119	39	30	20.8	0.54	17.91	686	305	93	569	10	579	84.4
4	12.20	0.91	13 3/8	18.08	0.23	18.45	0.39	338.7	119	39	30	20.8	0.54	17.91	686	378	93	560	10	570	83.2
5	12.30	1.562	16 7/8	26.23	0.23	26.0	0.67	338.12	117	41	30	21.4	0.56	25.44	974	548	93	790	10	800	82.2
6	12.35	1.562	16 7/8	26.23	0.23	26.0	0.67	338.12	117	41	30	21.4	0.56	25.44	974	565	93	815	10	825	84.7

The Pelton Water Wheel Company—San Francisco and New York

For description of plant, see page 107

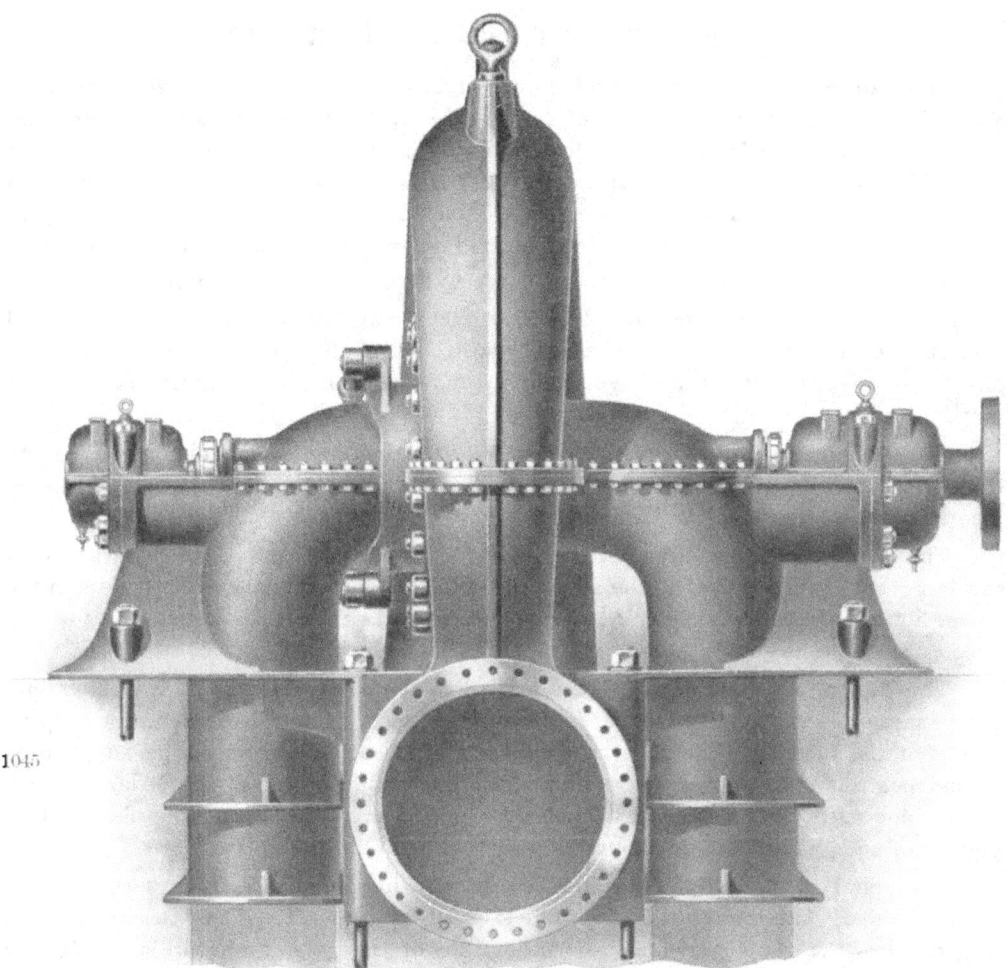

Typical Illustration of Double Pelton-Francis Spiral-case Turbine

The Claremont Power Company

The original plant of this company is located at Claremont, New Hampshire, and consists of two 400 horse-power turbines, and a steam plant of like capacity as a reserve. The growing business of the company called for a large increase in power, and to supply this a water right, located near Cavendish, Vermont, was secured.

The water power development consists of a concrete dam 36 feet high and about 250 feet in length, from which the water is led by means of a tunnel 8 feet in diameter and 300 feet long. To the lower end of the tunnel is connected a riveted steel pipe 6½ feet in diameter and 1150 feet in length, this terminating in a 7-foot diameter receiver at the power house. 80 feet distant from the power house is located a standpipe 100 feet high and 6 feet in diameter.

The plant consists of three 750 horse-power double 24-inch PELTON-FRANCIS turbines enclosed in spiral cast-iron cases, each turbine being direct-connected to a 450 k.-w. generator running at 600 r. p. m., the head being 120 feet. Each turbine is provided with a self-contained PELTON oil pressure governor and fly-wheel coupling of about 4000 pounds weight. There is one exciter unit of 50 k.-w. capacity running at 1000 r. p. m.

The total capacity of the plant is 2300 horse-power, and this is transmitted to Claremont, the two plants being run in synchronism. The power is used for operating an electric railroad and electric lighting system, in addition to which some 1500 horse-power is supplied to various manufacturing industries in that vicinity. The complete pipe line, as well as the turbines and governors, was furnished and installed by this Company.

The Schenectady Power Company

The plant of this company is situated on the Hoosic River near Schaghticoke, New York, about 15 miles from the City of Troy. The water power is obtained by throwing a low dam across the Hoosic River, thus diverting its flow into an open canal 2300 feet long. The canal has a carrying capacity of 1500 cubic feet per second, and terminates in a forebay, from which a 12½-foot steel pipe about 1000 feet long leads to a standpipe 40 feet in diameter and 50 feet in height, located on the hillside above the power house.

The installation consists of four special 52-inch *vertical* shaft PELTON-FRANCIS turbines, each direct-connected to a 3000 k.-w., 300 r. p. m., 40-cycle, three-phase General Electric generator, and two 250 horse-power horizontal shaft PELTON-FRANCIS turbines, direct-connected to 150 k.-w., 250-volt, 600 r. p. m., generators, for excitation of the main units.

Each main turbine unit is supplied with a 6-foot steel pipe 300 feet in length, leading from the standpipe to the turbine. The two exciter units are supplied by one 24-inch pipe line terminating in a "Y" branch.

The average effective head on the main turbines is 147 feet. The gate valves for the large units are located at the standpipe, and the valves for the exciter units in the power house. A sectional elevation of one turbine and generator is shown on page 112. Also on page 111 is shown a plan view of the turbine.

The design, it will be noted, bears a marked resemblance to a steam turbine unit, the generator being mounted on a distance ring, which, in turn, is carried on the spiral casing of the turbine. The governors are of the PELTON oil pressure type, supplied by an independent oil pumping system, and each is mounted on the distance ring between the turbine gates and generator. Each governor is capable of exerting 30,000 foot pounds of energy, but 7500 foot pounds is ample to handle the gates.

The spiral casings of the turbines are of cast-iron, made in four sections with flanged joints. The runners are of gun metal, cast in one piece. The gates are made of forged steel, the shaft projecting on either side of the gates and working in bronze bushed bearings. The water passage from the spiral casing into the runner has removable gun metal liner rings. The turbine runner is bolted to a coupling forged on the end of the generator shaft; this shaft is supported by a roller thrust bearing located on the top of the generator and running in oil, there being two "steady-bearings"—one directly under the thrust bearing end of the generator, and the other above the ring cover on the upper side of the turbine.

Immediately above the lower steady-bearing there is a small gear pump driven from the generator shaft, which supplies oil to the generator thrust and steady-bearings as well as to the steady-bearing of the turbine. The oil is collected in a reservoir below the steady-bearing, from whence it is pumped by the small gear pump above referred to and used over again. Special arrangements are provided to take care of any leakage from the turbine cases where the shafts project through, by means of drain pipes and ejectors.

The total weight of the revolving parts of each turbine and generator is approximately 42,000 pounds, and the design of the runner is such that when operating at full normal load about 90 per cent of this weight is taken care of by the upward water pressure exerted under the lower rim of the turbine, and by the reaction due to discharge into draft tube. Each main unit is provided with a mechanical hand-operated brake, enabling it to be brought to a full stop within about three minutes.

The oil pressure for operating the main turbine is taken from a central pumping system supplying oil to the four main turbine governors. This consists of a special PELTON duplex oil pump mounted on the side of the pressure tank and driven by a silent chain from a variable-speed electric motor.

The governors of the exciter units are of the PELTON self-contained oil pressure type, and are independent of the central pumping system.

The power from this plant, aggregating in excess of 20,000 horse-power, is transmitted twenty miles to Schenectady, New York, where it is used for running the immense works of the General Electric Company.

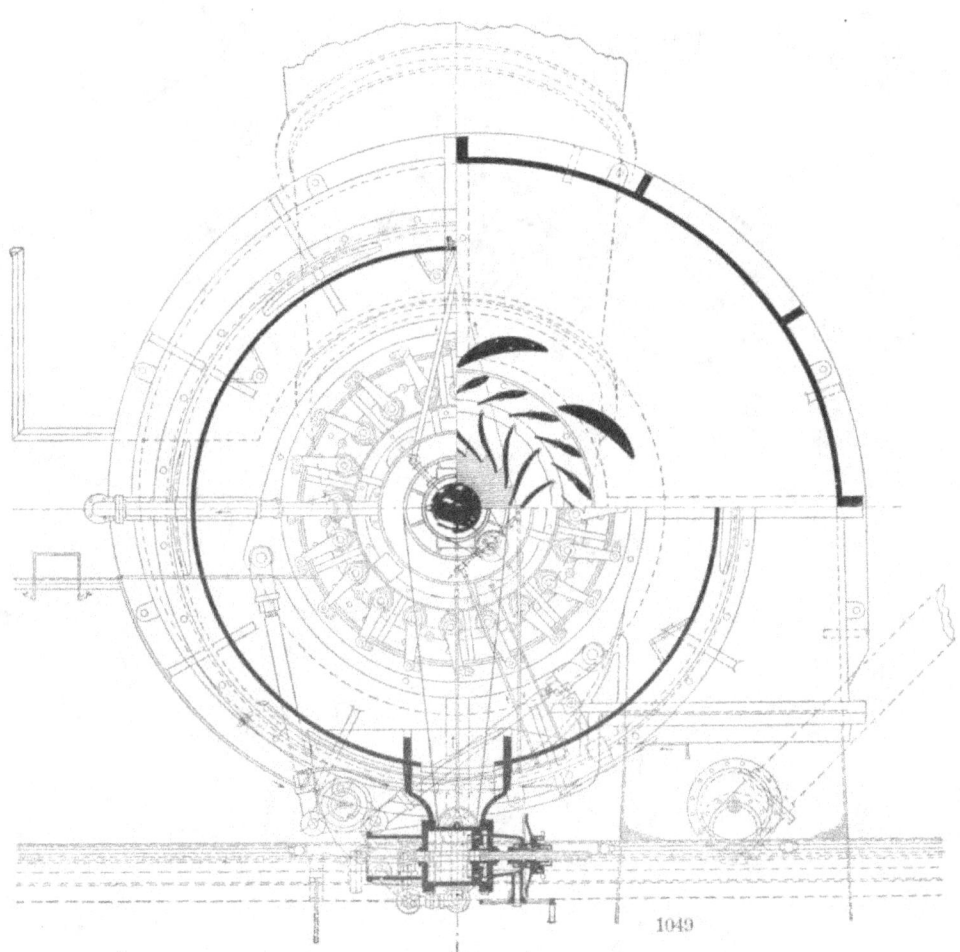

5000 Horse-power Vertical Shaft Pelton-Francis Turbine—Plan View
(See description above)

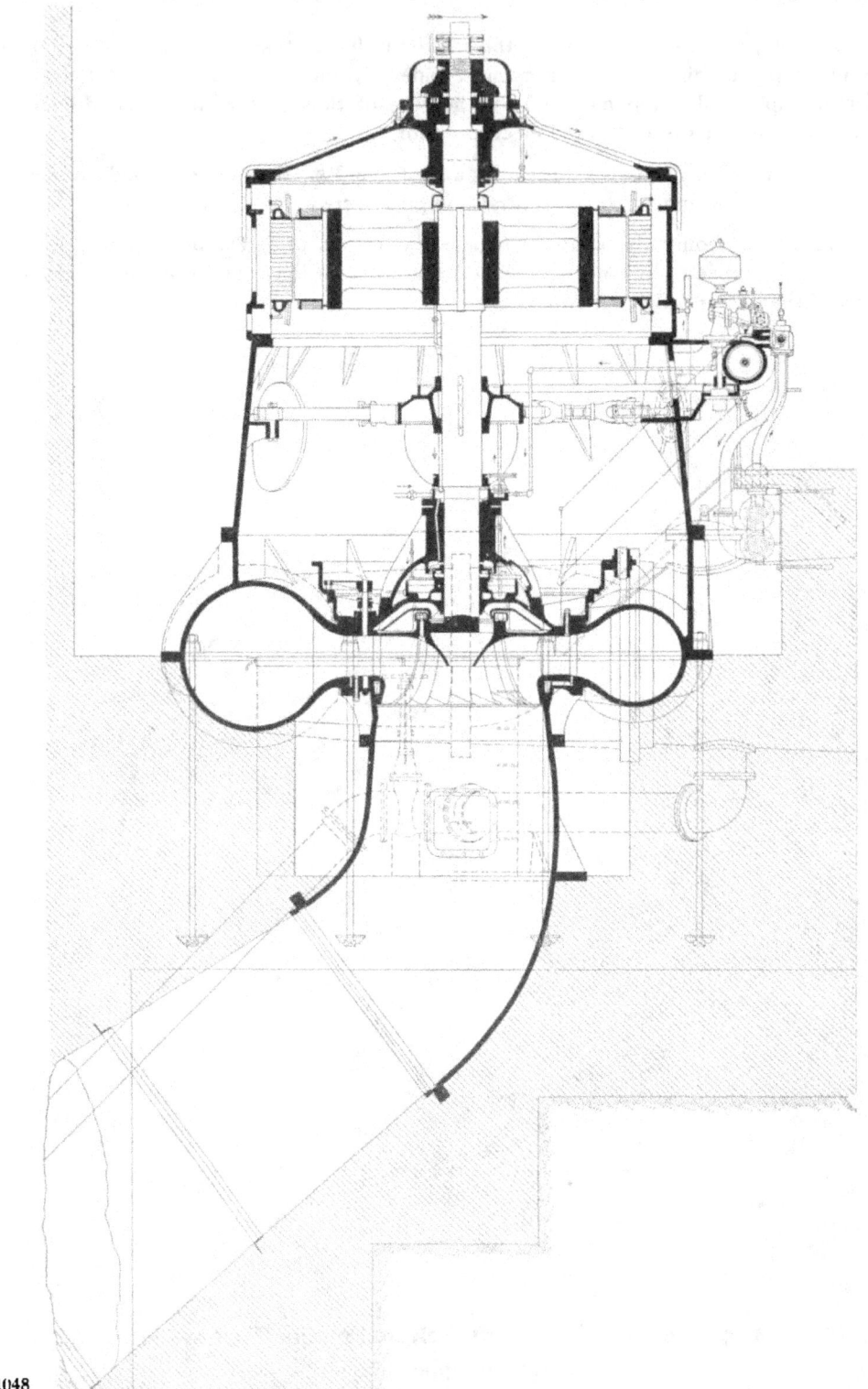

1048

Sectional Elevation 5000 Horse-power Vertical Shaft Pelton-Francis Turbine
(See description on page 110)

Pelton Telegraphic Code

Standard Wheels — Standard Motors

3-foot wheel, complete . . . . Abbais	6-inch motor . . . . . . Acment
4-foot wheel, complete . . . . Abcant	12-inch motor . . . . . . Acolite
5-foot wheel, complete . . . . Abvert	15-inch motor . . . . . . Adelton
6-foot wheel, complete . . . . Actran	18-inch motor . . . . . . Adlent
	24-inch motor . . . . . . Admont

Multiple Nozzle Wheels

2-nozzle 3-foot wheel, complete . . Baclen	18-inch quintex wheel, complete . . Bajan
2-nozzle 4-foot wheel, complete . . Baden	24-inch quintex wheel, complete . . Bants
2-nozzle 5-foot wheel, complete . . Badger	36-inch quintex wheel, complete . . Baque
2-nozzle 6-foot wheel, complete . . Baird	42-inch quintex wheel, complete . . Bardin
	48-inch quintex wheel, complete . . Baston

General Information

What sized wheel is advised for —— horse-power, under —— head? . . . . Caapon
Example : 50 horse-power is wanted under 150 feet head ; the inquiry would read thus :
    Caapon Ladrillo Maltine. [Codes of horse-power and heads are given on page 115].
The reply would be, Abcant, meaning that a 4-foot wheel is advised to fill above.
The sized wheel you order will not give the power wanted under conditions named . . Cabazon
You require more water or a higher head to get power wanted . . . . . Cadamas
There is not the power called for in the water under head given . . . . . Cadstrom
Standard wheel not suited to your requirements ; advise special wheel . . . . Cadogan
Data given not sufficient to make an intelligent estimate. Give particulars as to head, power
    required, amount of water, length and diameter of pipe line . . . . . . Calabon
We advise a deflecting nozzle with governor . . . . Camartic
We advise a deflecting nozzle with lever for hand regulation . . . . . . Cambosh
Give as accurately as possible the horse-power wanted . . . . Camjore
Await further instructions by letter . . . . Cameron
Recommend —— foot wheel for wood frame mounting . . . . . . Cameta
Speed —— revolutions per minute . . . . Camguin
What is the character and speed of driven machinery ? . . . . . . Camloch
We understand the pressure stated is effective at wheel . . . . . . Cammack
Horse-power available after deducting losses in pipe line . . . . . .
Wheel runner weighs —— pounds . . . . Camolin
Consider wheel should be —— horse-power capacity for —— per cent generator overload . Campalt
Wheel is capable of carrying —— per cent overload . . . . . . Canau
Wheel unit designed for maximum efficiency when developing sufficient power for normal output
    from generator, but capable of developing —— per cent overload . . . . Canavan
Advise —— with outboard bearing and sole plate . . . . Cancao
Iron-mounted wheel unit complete similar to illustration on page —— of catalog . . Canchan
What is k.-w. capacity and speed of generator? . . . . Candar
Generator speed is —— revolutions per minute . . . . Candler
Speed is too high for direct connection . . . . Canehill
Generator is to be of engine type . . . . Canerige
Generator is to be of two-bearing water wheel type . . . . . . Canfranc
Capacity —— k.-w. . . . . Cangal
Water wheel diameter approximately —— inches . . . . Canglas
Recommend wheel for mounting direct on compressor shaft . . . . . Canigou
What is the speed of compressor? . . . . Canister
What is the weight and diameter of compressor band wheel? . . . . . Canjilon
Water wheel is designed of sufficient weight to serve as fly-wheel . . . . . Canmore
Wheel for direct connection to compressor similar to engraving on page —— of catalog . Cannock
What should be the weight of water wheel for fly-wheel effect? . . . . . Candas
Pelton-Francis turbine . . . . Canopic
Complete Pelton-Francis turbine with draft tube and coupling . . . . . Canosa
Fly-wheel included . . . . Cantana
Mechanically-operated relief valve included . . . . Cantey
Telegraph promptly . . . . Capac
Mail promptly . . . . Capez
Would advise using —— wheels . . . . Capistran
Letter with full data and plans will be sent . . . . Carmentos
We are preparing plans and will forward as soon as completed . . . . . Cayugon
Advise to fill conditions named . . . . Chaparral
How soon and on what terms can you send a competent engineer to make examination of water
    power and estimate cost of plant? Wire answer . . . . Chidalgo
We will send you a competent engineer to make a survey and report on your water projects, to
    place designated, for —— dollars per day and expenses . . . . . . Cholone
Send at once one of your most experienced engineers to place designated in our last, on terms
    named, requesting him to report, on arrival, to —— . . . . . . Chromite
Our engineers are all engaged ; will endeavor to send one in about —— days. Will that answer? Champos

Relating to Prices

What will be the price of ——?	Dacton
Wire price f. o. b., San Francisco, of ——	Daices
Wire price f. o. b., New York, of ——	Danher
Price satisfactory; go ahead with order; shipping instructions will be mailed	Darrah
Price satisfactory; but do nothing until you receive plans and orders by mail	Dathol
The price named is net cash	Daylors
Price complete including foundation and frame bolts, but without pulley or wood frame	Deacon
Price f. o. b., New York	Dealton
Price f. o. b., San Francisco	Deanwood
Price f. o. b., ——	Deapney
Price c. i. f. ——	Deberke
Hand-controlled needle regulating nozzle, $ —— additional	Debessi
Combination needle and stream deflecting nozzle, $ —— additional	Deblois
Automatic needle nozzle, $ —— additional	Demoin
Price includes Pelton mechanical governor with all driving and operating connections	Decants
Price includes Pelton self-contained oil pressure governor with all driving and operating connections	Decherd
Iron-mounted wheel unit for direct connection to generator	Dedrick
Price includes shaft and bearings for engine-type alternator	Defunak
Add —— cents per pound for any additional weight required in water wheel	Degsdorf
Prices quoted are a close approximation; final figures when you are ready to close contract	Dektrip

Relating to Orders

Enter our order for wheel to develop —— horse-power under —— feet head	Fairfax
Example: Enter our order for wheel to develop 140 horse-power under 180 feet head; which would read: Fairfax Lancade Mancos.	
Enter order for wheel under —— feet head with water supply of —— cubic feet per minute, maximum	Falerno
Make in sections for mule transportation, no piece to exceed 250 to 300 pounds	Faleston
Execute this order with all possible dispatch	Falkner
Wire us when ready for shipment	Falstaff
Send —— extra buckets	Farslip
Send an old bucket from wheel as a sample	Faucher
Make shaft —— inches in diameter exact	Feglins
Keyway —— wide by —— deep	Fentop
Send gage for diameter of shaft	Ferguson
Send gage for bore of wheel	Fernando
Shall we furnish driving pulley? If so, give diameter and face	Fernside
Send driving pulley for wheel shaft of proper size to run counter with pulley —— diameter x —— face at —— revolutions	Filtons
Will belt direct from wheel shaft; driving pulley on wheel shaft must be —— diameter x —— face	Finchs
Send driving pulley —— x ——	Findley
Give speed of counter and diameter of pulley on same	Finnell
Give maximum power wheel is to develop	Firbunt
Send gage for wheel shaft	Fleener
Send sole plates and foundation bolts at once	Flonest
Furnish deflecting nozzle with rock-shaft, levers and quadrant without governor	Fowton
Wheel shaft should be extended to take driving pulley on each side of wheel	Foxen
Place pulley on right-hand side when facing wheel from nozzle side	Fredcam
Place pulley on left-hand side when facing wheel from nozzle side	Frilstop
Furnish one additional bearing	Frisbies
Provide wheel with ring-oiling journals	Frohms
Ring-oiling journals will cost —— extra	Fyffes

Relating to Payment

Enter order; remittance made through ——	Gablain
Draw on us, with bill of lading attached, through bank	Galions
Draw on us, with bill of lading attached, through house of ——.	Gantoin
Must establish credit here for amount of order through bank against documents	Gaviota
Remit for —— per cent of order; will draw for balance against documents	Gelston
You must put us in funds for —— per cent of amount of order before we can go on with it; will draw against documents for balance	Genevra

Relating to Regulation

Send style of governor you consider best adapted to our purpose	Habbard
Send governor with deflecting nozzle	Hafford
Governor is necessary for your purpose	Haywood
Governor is not required for your purpose	Heinfen
In connection with governor, you will require a fly-wheel approximately—— pounds in weight	Hinton
State character of load and probable variation	Hoaglin
Send —— governor	Hoamus
Pelton self-contained oil pressure governor	Hoback
Pelton special type governor with independent oil pump	Hockton
Combined needle deflecting nozzle	Hodnan

Relating to Water Supply

Cubic feet per minute	Iacua
We have a maximum supply of —— cubic feet per minute	Ibex
We have a minimum supply of —— cubic feet per minute	Idria
Supply variable, but can count on an average of —— cubic feet per minute	Igerna
What is your maximum supply of water in cubic feet per minute?	Ignasco
We cannot obtain more water than the amount stated	Indeek
If you cannot obtain more water, can the head be increased? If so, how much?	Iponia
The head can be increased to ——	Ipswich

Relating to Shipment

Await shipping instructions	Jaegers
Must be ready for shipment on the ——	Jaguas
Ship on steamer of the ——	Jalama
Ship on next steamer	Jamsan
Forward by fast freight	Jamul
Shipping weight will approximate —— pounds	Jandor
Shipment will approximate —— tons	Jandice
Forward by express	Janes
What is the earliest date on which you can ship?	Jaqua
On what date was your shipment made?	Jasmin
Shall we insure your shipment?	Jelleys
Insure our shipment for $ ——	Jempic
It will require about —— days to fill your order	Jepsum
Quote best freight rate to ——	Jewetta
Car-load rate to —— is ——	Jolont
Less car-load rate to —— is ——	Juapon
Rate to —— is $ —— per ton, weight or measure	Juston

Horse-power 10 to 2000

10	Labrea	150	Leavits	290	Liberty
20	Lacana	160	Lecuya	300	Lidells
30	Lacosta	170	Lefrancs	320	Liegan
40	Laddson	180	Leighton	340	Lillis
50	Ladrillo	190	Leland	360	Limas
60	Lagona	200	Lempton	380	Lincos
70	Lagracia	210	Lenstrom	400	Lindale
80	Lahonda	220	Leonas	420	Lindross
90	Lairds	230	Lerichs	440	Lisbon
100	Lajolla	240	Lethent	460	Lithane
110	Lachorbo	250	Levpont	480	Litton
120	Laporto	260	Lewiston	500	Livement
130	Lamande	270	Lexmont	1000	Livsome
140	Lancade	280	Leysant	2000	Lizcont

Head of Water in Feet

20-foot head	Mabels	230-foot head	Meacham	580-foot head	Midson
30-foot head	Machos	240-foot head	Meadins	600-foot head	Milpate
40-foot head	Maclays	250-foot head	Medias	620-foot head	Milford
50-foot head	Macus	260-foot head	Meehan	640-foot head	Milbrae
60-foot head	Madler	270-foot head	Meekers	660-foot head	Milner
70-foot head	Madera	280-foot head	Meineck	680-foot head	Milsaps
80-foot head	Madison	290-foot head	Melitta	700-foot head	Mindem
90-foot head	Mafost	300-foot head	Melrose	720-foot head	Mineota
100-foot head	Mahews	320-foot head	Mentone	740-foot head	Minturn
110-foot head	Malaga	340-foot head	Meraza	760-foot head	Miramar
120-foot head	Malakoff	360-foot head	Mercury	780-foot head	Mockton
130-foot head	Malkonus	380-foot head	Merdout	800-foot head	Mojeska
140-foot head	Malstown	400-foot head	Merigan	820-foot head	Mojave
150-foot head	Maltine	420-foot head	Merles	840-foot head	Moneta
160-foot head	Mampos	440-foot head	Merton	860-foot head	Montalvo
170-foot head	Manstrap	460-foot head	Mesant	880-foot head	Moraga
180-foot head	Mancos	480-foot head	Mesick	900-foot head	Morena
190-foot head	Manvil	500-foot head	Mesila	925-foot head	Morley
200-foot head	Marcel	520-foot head	Mesmer	950-foot head	Moulton
210-foot head	Marcuse	540-foot head	Mesquit	975-foot head	Mowrys
220-foot head	Maskot	560-foot head	Metsom	1000-foot head	Munchton

Relating to Pipe Line

What diameter pipe will be required for ——— horse-power under ——— head, the length being ——— feet?	Pachapa
What is the length of pipe line?	Pacheco
What is the diameter of pipe?	Pacolma
Is your pipe line already laid?	Paguay
Your pipe is too small to carry the required amount of water	Pahute
Advise consulting us in regard to pipe	Pajaro
What thickness of steel is advised for ——— diameter pipe under ——— feet head?	Palmos
Slip-joint pipe	Panoche
Collar- and sleeve-joint pipe	Parais
Flanged pipe	Pasiana
Wrought-iron welded pipe with flanged joints	Pawneck
Send ——— feet of pipe of proper gage and diameter to fill our requirements	Penrose
For approximate estimate of cost of pipe, see price list in catalog	Perrys
Must have a profile of pipe line in order to make an intelligent estimate of size, weight and cost of pipe	Pescados
——— gage pipe is too light to stand the pressure; advise ——— gage	Petrolia
Advise using ——— feet each, number ——— and ——— steel	Phoenix
Advise pipe of several diameters nested to save freight; shall we make a change?	Pidston
Your pipe should be made of ———gage steel	Piedro
No. 18 U. S. standard wire gage	Pinacate
No. 16 U. S. standard wire gage	Pinebaugh
No. 14 U. S. standard wire gage	Placentia
No. 12 U. S. standard wire gage	Pleyton
No. 10 U. S. standard wire gage	Poormas
No. 8 U. S. standard wire gage	Popkins
No. 7 U. S. standard wire gage	Potsdam
No. 6 U. S. standard wire gage	Poways
Send pipe cut, punched and formed for mule packing	Preatos
Length of section of pipe must not exceed ——— feet	Presidio
Make pipe in ——— sizes to facilitate shipment	Prieton
——— pipe will weigh ——— pounds per foot	Proberta
——— inch pipe, No. ———	Pulcurate

Relating to Length of Pipe

5 feet	Sabloa	200 feet	Semblos	2500 feet	Sibinto
10 feet	Sabinal	250 feet	Sebranto	3000 feet	Sicorno
20 feet	Sablato	300 feet	Secreto	3500 feet	Sidesto
30 feet	Sabuyla	400 feet	Seculto	4000 feet	Sigrado
40 feet	Sabinto	500 feet	Sedono	4500 feet	Sijando
50 feet	Sapuisco	600 feet	Sedroma	5000 feet	Silhino
60 feet	Sahenta	700 feet	Sefranca	5500 feet	Simona
70 feet	Saspico	800 feet	Segona	6000 feet	Sintosa
80 feet	Satrona	900 feet	Seguro	7000 feet	Siperta
90 feet	Sareilo	1000 feet	Sejaron	8000 feet	Sirosis
100 feet	Sanpete	1500 feet	Sejinto	9000 feet	Sistema
150 feet	Salesco	2000 feet	Serina	10000 feet	Sitasco

Relating to Diameter of Pipe

1-inch diameter	Tacita	11-inch diameter	Tanka	26-inch diameter	Tecian
2-inch diameter	Tacuba	12-inch diameter	Tanque	28-inch diameter	Teloan
3-inch diameter	Tacpeto	13-inch diameter	Tapias	30-inch diameter	Teltos
4-inch diameter	Talaca	14-inch diameter	Tasco	32-inch diameter	Tecpan
5-inch diameter	Talcoch	15-inch diameter	Teabo	34-inch diameter	Tecuala
6-inch diameter	Talpa	16-inch diameter	Tepon	36-inch diameter	Tejano
7-inch diameter	Tamah	18-inch diameter	Teculi	38-inch diameter	Tejeria
8-inch diameter	Tampico	20-inch diameter	Tecuma	40-inch diameter	Temaxo
9-inch diameter	Tamos	22-inch diameter	Tecute	44-inch diameter	Temoson
10-inch diameter	Tamuin	24-inch diameter	Tecax	48-inch diameter	Tempoal

NOTE.—The above supersedes all previous codes, and can be used in conjunction with the A B C and Lieber's codes.

www.ingramcontent.com/pod-product-compliance
Lightning Source LLC
Chambersburg PA
CBHW081133170526
45165CB00008B/2656